SO-BFD-691

ORGANIC SYNTHESES

ORGANIC SYNTHESES

AN ANNUAL PUBLICATION OF SATISFACTORY METHODS FOR THE PREPARATION OF ORGANIC CHEMICALS

VOLUME 58

1978

BOARD OF EDITORS

JOHN WILEY AND SONS
NEW YORK · CHICHESTER · BRISBANE · TORONTO

Published by John Wiley & Sons, Inc.

Library of Congress Catalog Card Number: 21-17747
ISBN 0-471-04739-2

Printed in the United States of America

10 9 8 7 6 5 4 3 2 1

NOMENCLATURE

Both common and systematic names of compounds are used throughout this volume, depending on which the Editor-in-Chief feels is more appropriate. Preparations appear in alphabetical order of names of the synthetic procedures. To group preparations of common interest, titles have been adapted for appropriate alphabetical listing. The *Chemical Abstracts* indexing name for each title compound, if it differs from the title name, is given as a subtitle. Systematic *Chemical Abstracts* Nomenclature is provided for both the 8th and 9th Collective Indexes in an appendix at the end of each preparation. Registry numbers, which are useful in computer searching and identification, are also provided in these appendixes. Whenever two names are concurrently in use, and one name is the correct *Chemical Abstracts* name, that name is adopted. For example, both diethyl ether and ethyl ether are normally used. Since ethyl ether is the established *Chemical Abstracts* name for the 8th Collective Index, it has been used in this volume. The 9th Collective Index name is 1,1'-oxybisethane which the Editors consider too cumbersome. The prefix *n*- is deleted from *n*-alkanes and *n*-alkyls. All reported dimensions are now expressed in Système International units.

SUBMISSION OF PREPARATIONS

Chemists are invited to submit for publication in *Organic Syntheses* procedures for the preparation of compounds that are of general interest, as well as procedures that illustrate synthetic methods of general utility. It is fundamental to the usefulness of *Organic Syntheses* that submitted procedures represent optimum conditions, and the procedures should be checked carefully by the submitters, not only for yield and physical properties of the products, but also for any hazards that may be involved. Full details of all manipulations should be described, and the range of yield should be reported rather than the maximum yield obtainable by an operator who has had considerable experience with the preparation. For each solid product the melting-point range should be reported, and for each liquid product the boiling-point range and refractive index should be included. In most instances it is desirable to include additional physical properties of the product, such as ultraviolet, infrared, mass, or nuclear magnetic resonance spectra, and criteria of purity such as gas chromatographic data. In the event that any of the reactants are not commercially available at reasonable cost, their preparation should be described in as complete detail and in the same manner as the preparation of the product of major interest. The sources of the reactants should be described in the Notes section, and physical properties such as boiling point, index of refraction, and melting point of the reactants should be included except where standard commercial grades are specified.

Beginning with Volume 49, Methods of Preparation (Sec. 3) and Merits of the Preparation (Sec. 4) have been combined into Discussion (Sec. 3). This section should include descriptions of related and practical methods. Other published methods that have no practical synthetic value do not need to be mentioned. Those features of the procedure that recommend it for publication in Organic Syntheses should be cited (synthetic method of considerable scope, specific compound of interest not likely to be made available commercially, method that gives better yield or is less laborious than other methods, etc.). If possible, a brief discussion of the scope and limitations of the procedure as applied to other examples, as well as a comparison of the particular method with the other methods cited, should be included. If necessary to the understanding or use of the method for related syntheses, a brief discussion of the mechanism may be placed in this section. The present emphasis of Organic Syntheses is on model procedures rather than on specific compounds (although the latter are still welcomed), and the Discussion should be written to help readers decide on the value of the procedure in their research. Three copies of each procedure should be submitted to the Secretary of the Editorial Board. An accompanying letter setting forth the features of the preparations that are of interest or value is helpful to the Board.

Additions, corrections, and improvements to the preparations previously published are welcomed; these should be directed to the Secretary.

JAMES BRYANT CONANT

March 26, 1893–February 11, 1978

James Bryant Conant, chemist, educator, administrator, states-
man, was a major architect of, and participant in, the phenomenal
increase in the power, prestige, and accomplishments of U.S.
science that took place during the first third of this century. Prior
to his time, American science and American universities suffered
by comparison with their European counterparts; during the decades just preceding World War II, however, both U.S. science and
the universities—with considerable help from Conant—pulled
themselves "up by their bootstraps." U.S. chemistry grew to international stature, partly through the initiation of the series *Organic
Syntheses*. Conant strongly supported Roger Adams in his proposal
for a regular publication devoted to reliable synthetic procedures;
he was a member of the first Board of Editors and served as Editor-
in-Chief of Volumes 2 and 9.

His death on February 11, 1978, at the age of 84, brought to a
close his remarkable career (or, as he preferred to regard it,
careers). His achievements as a chemist were followed by his
creative administration as President of Harvard University from
1933–1953, and by his success as U.S. High Commissioner to
Germany (1953–1955) and Ambassador to the Federal Republic
of Germany (1955–1957). While on leave from the Presidency of
Harvard, he served with distinction as Chairman of the National
Defense Research Committee and Deputy Director of OSRD
(1941–1946); subsequent to his assignment as ambassador, he
undertook a massive study of U.S. secondary education; his books
on that subject ("The American High School Today," "The
Citadel of Learning," "Slums and Suburbs: A Commentary on
Schools in Metropolitan Areas," and many others) focused attention on a vital, but at that time somewhat neglected, part of our
society. Other obituaries undoubtedly will emphasize his roles as

administrator, diplomat, and educator, and record the many honorary degrees and other awards that he truly deserved and received. The chemical community remembers and honors him principally for his original contributions to chemistry, as well as what he did for science (in fact, for all scholarship) while President of Harvard.

Organic chemistry in 1920 was dominated by European scientists: Richard Willstätter, Leopold Ruzicka, Hans Fischer, Paul Karrer, Heinrich Wieland, Aldolph Windaus, Hans Meerwein, Robert Robinson, Arthur Lapworth, and many others. The discipline, then as now, was concerned both with natural products and with fundamental theory, and Conant, almost alone among American chemists, was at the forefront of both major thrusts. Although he himself was most proud of his contribution to the determination of the structure of chlorophyll, his extraordinarily original innovations in physical-organic chemistry had a much greater impact on the development of science. The thirties and forties were marked by what amounted to a revolution in understanding of reaction mechanisms; Conant's ideas and discoveries gave strong impetus to progress in this area. In particular, he and N. F. Hall introduced the idea of "superacidity" for solutions of strong acids in nonaqueous solvents, and with G. M. Bramann, Conant illustrated his ideas by showing that both sodium acetate and perchloric acid catalyze the acetylation of β-naphthol in acetic acid; acid catalysis increased the rate a millionfold. Sodium acetate had, of course, often been used by synthetic chemists as a catalyst; the realization that it functions as a strong base in acetic acid was of fundamental importance. Similarly, Conant and G. W. Wheland (and, later W. K. McEwen) initiated the quantitative understanding of extremely weak acids such as cyclopentadiene, using triphenylmethyl sodium in ether as the needed strong base. These were germinal contributions to the theory of nonaqueous solutions.

Furthermore, Conant originated, or helped originate, several other fundamental aspects of chemistry. He was among the group who applied ^{11}C (it was the only isotope of carbon then available) to a trail-blazing study of a metabolic pathway. He and G. B. Kistiakowsky initiated the measurement of the heats of hydrogenation of organic compounds, so as to improve the precision of thermodynamic data relative to those available from heats of combustion. He and P. W. Bridgman were the first to investigate the effects of extremely high pressure on the rates of reaction of organic compounds; they discovered the acceleration of polymerization by pressure. Conant's investigations (with L. F. Fieser, among others) of the reversible oxidation–reduction potentials of quinones, and his studies of irreversible electrochemical oxidations and reductions, were far ahead of their time. In his work with P. D. Bartlett on the mechanism of semicarbazone formation, he distinguished sharply between kinetic and thermodynamic control; that paper alone exerted a powerful influence on developing theory.

One interesting aspect of Conant's work in the biochemical area was the discovery that copper is the essential metal of the prosthetic group in hemocyanin, the oxygen-carrying pigment of crustaceans. In his major biochemical work, Conant and his collaborators discovered the role of autoxidation in the so-called "phase test" for chlorophyll—a complex and previously confusing series of reactions initiated by strong base. Hans Fischer and his many collaborators in Munich were the chief investigators of chlorophyll, as they had been for hemin; in 1935 Fischer suggested an essentially correct solution to the structural problem. Although Conant's part in the chlorophyll story was a minor one, one can only speculate as to what he might have accomplished had he not been appointed in 1933 to the Presidency of Harvard, but had continued in chemistry. The same speculation holds for his investigations of physical-organic chemistry, since he left the field just as it was beginning its exponential growth phase.

Conant's students have built on the ideas he introduced, both in physical-organic chemistry and in biochemistry. His co-workers included L. F. Fieser, P. D. Bartlett, and me, all of whom (after intervals elsewhere) returned to the Harvard Chemistry Department, and A. M. Pappenheimer, who was appointed in the Biology Department; they also included G. W. Wheland at the University of Chicago, Alsoph Corwin at Hopkins, Robert Lutz at Virginia, Emma Dietz Stecher at Barnard, Jack Astin at Penn State, and several others whose research advanced both mechanistic and biological chemistry. In addition to research, Conant and his collaborators advanced the teaching of chemistry through their writings, including especially "The Chemistry of Organic Compounds," a highly innovative and successful basic textbook, published jointly with his scientific co-worker, A. H. Blatt. Conant gave direction to these works, and through his example to a whole generation of U.S. chemists, the generation that came up to European (and his) standards.

Conant was important not only to chemistry and to science, but also to scholarship through being President of Harvard. An activist president, he introduced scholarships for needy students, promoted geographical distribution in the college, and emphasized general education. But those who see these innovations as his major contributions have missed the point of his presidency. The Emperor Augustus boasted that he found Rome a city of bricks and left it in marble; Conant found Harvard a college and left it a university. In large part the transformation rested on Conant's introduction of the *ad hoc* committees that apply rigid standards to each tenure appointment. But perhaps even more important was the obvious intent of the system. Conant himself was a scholar who loved and honored research. The measures he introduced established research—and therefore teaching at the graduate level—as comparable in importance with undergraduate instruction. And because Harvard was a prestigious institution (and, too, because the time was ripe for such leadership), this upgrading of the graduate school at Harvard served as an example to many other universities.

Conant was able to devote his full effort to Harvard for only a few years; as the Nazi military threat came to dominate the world, Conant became increasingly involved with national affairs. As Chairman of the National Defense Research Committee, he provided effective scientific leadership during World War II. After the war he was appointed the first Chairman of the National Science Board, and in that role helped to initiate the policies that proved so successful in encouraging the development of science, especially in the U.S., but also abroad.

Conant, then, was perhaps the individual most instrumental in bringing the level of U.S. scholarship in general, and of chemistry in particular, up to, and in many cases beyond, the standards previously set in Oxford, Munich, and Zurich. His great achievement was not the solution of some particular problem in reaction mechanisms, or some specific determination of structure, or some generalization with respect to organic chemistry—although he did contribute in all of these areas. His achievement was rather that he raised the level of science here and throughout the world, a monument truly worth having.

March 1978   F. H. WESTHEIMER

LOUIS F. FIESER

April 7, 1899–July 25, 1977

Louis Frederick Fieser, one of the most prolific contributors to *Organic Syntheses*, as well as a member of the Advisory Board and Editor-in-Chief of Volume 17, died on July 25, 1977, of pneumonia at his home in Belmont, Massachusetts. He is survived by his wife, Mary.

Born in Columbus, Ohio, on April 7, 1899, he was the son of Louis Frederick Fieser and Martha Victoria Kershaw Fieser. He attended Douglas School and East High in Columbus, and then Williams College, where he received the A.B. degree in 1920. Although his major was chemistry, he was also strongly attracted to English and Philosophy, and he was a member of the unbeaten varsity football team of 1919. He completed work for the Ph.D. degree at Harvard in three and one-half years, then continued working for a year and a half with his major professor, James Conant, on biochemical research. Immediately thereafter, in 1924, he spent a short time doing research in Germany, then at Oxford, before accepting a teaching position at Bryn Mawr College, where he remained an assistant and associate professor until 1930, when he was invited to join the Harvard faculty.

At this time Mary Peters, a member of his second class at Bryn Mawr, entered Radcliffe College to do graduate work under his direction. Her formal candidacy for the Ph.D. degree was brought to a close by their marriage in 1932; however, she continued to do research and was appointed Research Fellow of the Department of Chemistry.

In 1937 Fieser was promoted to full professor, and in 1939 he became the Sheldon Emery Professor of Organic Chemistry, a position he held until 1968, when he became Professor Emeritus.

During his early years as an academician Fieser became one of the world's leading experts in quinone chemistry, his most notable

xiii

achievement in this area being an elegant synthesis of Vitamin K_1 in 1939. He also developed an interest in the cancer problem, and by 1938 he had a large group of collaborators engaged in the synthesis of carcinogenic and related hydrocarbons in an effort to discover how these substances act. Out of this study came a facile synthesis of the potent carcinogen methylcholanthrene. He had occasion to prepare a comparison specimen of this substance with his own hands, according to the classical method, starting with the bile acid desoxycholic acid. Thus his interest in steroids was engendered, paving the way for a course, then a book, "Natural Products Related to Phenanthrene," which covered this exciting new field. He never lost interest in this area, and later, after two further editions of the phenanthrene book (the last jointly authored with Mary), Louis and Mary Fieser published the famous book entitled "Steroids," a classic in the field.

With the advent of World War II Fieser was obliged to suspend all previous research and teaching to work on problems for the National Defense Research Committee. Thus for three and one-half years he was involved with the development of new incendiaries (e.g., Napalm), antimalarials, and syntheses of cortisone. He became especially interested in the antimalarial program, which included a large number of collaborators, and which yielded some extremely potent new substances in the naphthoquinone series.

Professor Fieser always had a very strong interest in teaching, and his dynamic and vibrant personality, along with his original and colorful style of lecturing in the elementary organic course, served to communicate to students his love of science and the pleasure to be derived from a well-executed experiment. Many of his former students remember this course as a great influence on their scientific careers. From his teaching evolved the widely adopted textbook "Organic Chemistry," written in collaboration with his wife, which went through several editions and was translated into eight foreign languages. An expanded version of this book was called "Advanced Organic Chemistry."

The Fiesers proved to be an extraordinarily effective writing team, producing some of the most important scientific books of the century. Other Fieser and Fieser books include "Style Guide for Chemists" (1960), "Topics in Organic Chemistry" (1963), "Current Topics in Organic Chemistry" (1964), and "Reagents for

Organic Syntheses," Vol. 1 (1967), Vol. 2 (1970), Vol. 3 (1972), Vol. 4 (1974), Vol. 5 (1975), and Vol. 6 (1977). The "Reagents" series has proved to be extraordinarily important to the practicing synthetic chemist. Fortunately Mrs. Fieser has been continuing the work and has just sent Vol. 7 to the publisher.

As a practicing organic chemist who loved to work with his hands, Louis Fieser published 40 papers based on his own experimentation and, naturally, he took great interest in laboratory courses. He is famous for designing the Martius yellow sequence, named after the yellow dye that was the first of seven compounds obtained from 5 g of starting material; this became the basis of an annual prize competition in his class. When successful preparation of all seven compounds in 3–4 hours came to be regarded as a superior performance, Fieser began entering the competition each year along with the students. His best time was 1 hour and 59 minutes, the record for many years. He enjoyed devising new and interesting experiments for the students, and his laboratory manuals ("Experiments in Organic Chemistry" and "Organic Experiments"), published in six editions, were possibly the best available. In these, particularly, he communicates his skill as a virtuoso experimentalist.

In his constant search for better methods of teaching he made a 60-minute color-sound movie "Techniques of Organic Chemistry," and developed a set of precise plastic molecular models, which are larger than, but have the same relative dimensions as, Dreiding models. Unlike the latter, however, the Fieser models have been so inexpensive to manufacture that even undergraduate students have been able to afford a set.

Because of his expertise in carcinogens, Fieser was, in 1962, appointed by the Surgeon General to a committee to study the matter of the relationship of smoking to health. In 1964, after intensive study and deliberation, his committee submitted the now famous 387-page report expressing the unanimous opinion that "cigarette smoking is a health hazard of sufficient importance to warrant remedial action." Ironically Fieser, himself a chain smoker, had to undergo surgery for lung cancer the next year. He was very ill from emphysema and bronchitis too, and might not have survived had he not been endowed with an extraordinarily strong constitution. Having recovered, he was able to continue his productive writing career beyond his retirement from Harvard in 1968. In the spring semester of 1968 he was appointed Nielson Professor at Smith College.

Fieser published a total of 341 research papers, 36 of which with Mary, and over 20 books (the majority with Mary) which represent some of the very best writing found in science—clear, highly readable, elegant, and exciting.

He received an honorary D.Sc. degree from Williams in 1939 and was elected to the National Academy of Sciences in 1940. Other awards included the Katherine Berkham Judd prize for Cancer Research (1941), D. Pharm Honoris Causa, Université de Paris (1953), Manufacturing Chemists Association Award for Teaching (1959), Norris Award for Teaching (1959), and the William H. Nichols Medal (1963).

One of his most distinguished colleagues characterized Fieser as a man of extraordinary zeal, whose powerful and energetic constitution allowed him to transform that zeal into action. Not only was his colorful presence a feature of the Harvard scene for decades, but his influence through books and scientific contributions remains worldwide.

February 1978   WILLIAM S. JOHNSON

EDWARD P. HAMILTON

October 3, 1883–December 29, 1977

Edward P. Hamilton, retired President and Chairman of the Board of John Wiley & Sons, New York, and a long-time friend and supporter of *Organic Syntheses*, died December 29, 1977 in his ninety-fifth year.

Edward P. Hamilton was born in East Orange, New Jersey, son of the late Edward P. and Alice Wiley Hamilton. He attended the Hill School, Pottstown, Pennsylvania, and was graduated from Rensselaer Polytechnic Institute, Troy, New York, in 1907, with a degree in civil engineering. The next seven years were spent in sanitary, hydraulic, and construction projects in the Catskills, Berkshires, and in Cuba. In 1914 he joined the staff of John Wiley & Sons, the oldest book publishing house in New York City, established in 1807 by Mr. Hamilton's great-grandfather, Charles Wiley.

Mr. Hamilton was elected secretary of John Wiley & Sons in 1916, but, since he had joined Squadron A Cavalry of the New York National Guard, he was called to serve in the Mexican Border action under General John J. Pershing. During 1917–1918, Mr. Hamilton was a First Lieutenant, 306 Field Artillery, 77th Division. In France, he participated in the Baccarat Sector, and later in the Oise–Aisne and Meuse–Argonne Offensives. During a reconnaissance in the heart of the Argonne Forest in September of 1918, Lts. Hamilton and Salza were captured. Released after the November 11, 1918, Armistice, Lt. Hamilton was promoted to Captain in the Field Artillery Reserve in 1919. After discharge, he returned to work at Wiley and resumed visits to universities in search of scientific and technical books.

In 1920, during a visit to the University of Illinois (Urbana) Mr. Hamilton met Roger Adams.[1] They discussed the publication of a series of small annual volumes containing detailed checked preparations of organic compounds. As the result, the first volume of *Organic Syntheses* was published in 1921. This was a risky venture. There was no assurance that such specialized scientific books could be profitably sold; the country was in the depths of the post–World War I depression (the Dow Jones Industrials stock average was 65). However, the publication continued throughout the 1930–1946 Depression and continues today. Also, every 10 years a "Collective Volume" was published; now five Collective Volumes and a 50-year Cumulative Index have appeared.[2] Later, beginning in 1942, the *Organic Reactions* Series was published, and in 1949 *Biochemical Preparations* started. Mr. Hamilton's friendship led to the publication of the first two volumes of "Organic Chemistry; An Advanced Treatise," edited by Henry Gilman in 1938. All these publications by John Wiley & Sons helped the education of chemists who were advancing organic chemistry both industrially and scientifically.

Mr. Hamilton advanced in the Wiley organization, becoming Vice President and Treasurer in 1925, President in 1941, Chairman of the Board in 1956, and Honorary Chairman in 1966. Many fine college representatives, editors, and administrators were brought into the Wiley organization during his active years.

Mr. Hamilton was a long-time supporter of his alma mater, Rensselaer Polytechnic Institute, serving as an officer in the R.P.I. Alumni Association, a *Patroon* of R.P.I., and a member of the Board of Trustees for 20 years. In 1957, when John Wiley & Sons celebrated its 150th birthday,[3] R.P.I. awarded Mr. Hamilton an Honorary Doctor of Engineering Degree and elected him a life trustee. He established the Edward P. Hamilton foundation at R.P.I., which has supported many programs over the last 30 years. In 1976 he endowed the Edward P. Hamilton Chair for a Distinguished Educator, and in 1977 established a $1000 annual prize to future recipients of the Distinguished Faculty Award, now known as the William H. Wiley Award, in memory of his uncle who earned an engineering degree from R.P.I. in 1866.

Mr. Hamilton was a volunteer worker for more than 30 years in the Boys Club of New York City, and was President and a member of the Board of Trustees of the Kips Bay Boys Club, an organization serving about 2000 boys in an underprivileged area. He enjoyed horseback riding, spending many of his vacations on ranches in Wyoming and Montana, and had many cowboy friends. He was treasurer of a riding organization in Westchester County, New York.

Mr. Hamilton took an interest in international trade and after World War II traveled on behalf of a national organization of book publishers in Australia, New Zealand, the Philippines, Japan, and Latin American countries. He served on the Engineers Joint Council Committe on International Relations, and on the board of the Engineering Index. He was a member of *Theta Xi* and served as president for many years. He was elected to *Sigma Xi*, was a Life Member of the American Society of Civil Engineers, and was also a member of American Institute of Mining and Metallurgical Engineers, in addition to belonging to and supporting many clubs in New York City.

Mr. Hamilton's career as engineer turned publisher and as a philanthropist had a great influence on the advancement of science and technology, and thus on the social and economic progress that such advancements bring about. His many Boys Club members, cowboy friends, students and faculty at R.P.I., Editors of *Organic Syntheses*, colleagues at John Wiley & Sons, and many other friends, all are indebted to this modest but distinguished citizen.

RALPH L. SHRINER

March, 1978

1. Roger Adams, "Fifty Years of Organic Syntheses," *Org. Syn.* **50,** vii (1970).
2. R. L. Shriner and R. H. Shriner, "Organic Syntheses History," Appendix B, *Org. Syn. Cum. Indexes* (1976).
3. "The First One Hundred and Fifty Years. A History of John Wiley and Sons, Inc. (1807–1957)," New York, N.Y., 1957.

PREFACE

Through prudent selection, careful checking, and publication of procedures, *Organic Syntheses* has, since 1921, provided organic chemists with reliable experimental directions. The procedures appearing in this annual volume provide specific examples of important synthetic methods or precise directions for the preparation of intriguing compounds, starting materials, or reagents.

Dedication of annual volumes of *Organic Syntheses* to specific areas has often been suggested and considered. The present Editor-in-Chief had hoped to concentrate on organofluorine and heterocyclic chemistry in this volume. However, the process of obtaining and carefully checking preparations is not easily relegated to well-defined time intervals. The submitters often find that preparing an *Organic Syntheses* procedure is different, and often more difficult, than preparing a manuscript for publication in a journal. The process of checking a procedure can also lead to major delays if difficulties are encountered. The published procedures must be reproducible by an organic chemist with normal laboratory experience. These procedures are also often models that can be easily adapted for preparation of analogs or related compounds.

Considering these major factors, the active Board of Editors constantly revises the operating procedures for *Organic Syntheses* to better meet the current needs of organic chemists. An ongoing activity is to publish guides for submitters and checkers. A style guide for *Organic Syntheses* was prepared early in this decade and is available on request from the Secretary. A revision of this guide to conform with current ACS publication policies, including *Chemical Abstracts* Nomenclature, is in process, and a printed version of this guide will be available in the near future for distribution to potential submitters. The format recommended in this guide will apply to preparations to be included in the next decade, in Volumes 60–69. An abbreviated form of this guide for submitters should be published in Volumes 59 and 60.

In the current volume, 29 checked procedures are grouped on the basis of potential synthetic utility.

Heteroatom groups such as boron or silicon can activate or direct synthetic reactions. Use of such activation has become of major importance in organic syntheses. Examples in this volume are BORANES IN FUNCTIONALIZATION OF DIENES TO CYCLIC KETONES: BICYCLO[3.3.1]NONAN-9-ONE and BORANES IN FUNCTIONALIZATION OF OLEFINS TO AMINES: 3-PINANAMINE. Use of trimethylsilyl or trimethylsilyloxy groups to activate a 2-butenone or a butadiene are illustrated by the preparations 3-TRIMETHYLSILYL-3-BUTEN-2-ONE: A MICHAEL ACCEPTOR; 3-TRIMETHYLSILYL-3-BUTEN-2-ONE AS MICHAEL ACCEPTOR FOR CONJUGATE ADDITION ANNELATION: cis-4,4a,5,6,7,8-HEXAHYDRO - 4a,5 - DIMETHYL - 2(3H) - NAPHTHALENONE; and 2-TRIMETHYLSILYLOXY-1,3-BUTADIENE AS A REACTIVE DIENE: DIETHYL trans-4-TRIMETHYLSILYLOXY-4-CYCLOHEXENE-1,2-DICARBOXYLATE. Sulfur substitution also continues to be of high interest, and three preparations on sulfide synthesis are included: BENZYL SULFIDE; DIALKYL AND ALKYL ARYL SULFIDES: NEOPENTYL PHENYL SULFIDE; and UNSYMMETRICAL DIALKYL DISULFIDES: sec-BUTYL ISOPROPYL DISULFIDE.

Organometallic reagents and catalysts continue to be of considerable importance, as illustrated in several procedures: CARBENE GENERATION BY α-ELIMINATION WITH LITHIUM 2,2,6,6-TETRAMETHYLPIPERIDIDE: 1-ETHOXY-2-p-TOLYLCYCLOPROPANE: CATALYTIC OSMIUM TETROXIDE OXIDATION OF OLEFINS: PREPARATION OF cis-1,2-CYCLOHEXANEDIOL; COPPER CATALYZED ARYLATION OF β-DICARBONYL COMPOUNDS: 2-(1-ACETYL-2-OXOPROPYL)BENZOIC ACID; and PHOSPHINE-NICKEL COMPLEX CATALYZED CROSS-COUPLING OF GRIGNARD REAGENTS WITH ARYL AND ALKENYL HALIDES: 1,2-DIBUTYLBENZENE.

Halogen substitution continues to be of considerable value. A method illustrating the use of a new, versatile fluorinating reagent is given in FLUORINATIONS WITH PYRIDINIUM

POLYHYDROGEN FLUORIDE REAGENT: 1-FLUORO-
ADAMANTANE. Dichloroalkane synthesis is shown in cis-DI-
CHLOROALKANES FROM EPOXIDES: cis-1,2-DICHLORO-
CYCLOHEXANE. Nitrile functionality can be introduced
from a ketone, as in NITRILES FROM KETONES:
CYCLOHEXANENITRILE, or from a reactive diene, as shown
in 2,3-DICYANOBUTADIENE AS A REACTIVE INTER-
MEDIATE BY in situ GENERATION FROM 1,2-DICYANO
CYCLOBUTENE: 2,3-DICYANO-1,4,4a,9a-TETRAHYDRO-
FLUORENE.

Macrocyclic derivatives are of considerable importance in
biological areas and as complexing agents, particularly for metals.
Macrocyclic examples are given in MACROLIDES FROM
CYCLIZATION OF ω-BROMOCARBOXYLIC ACIDS: 11-
HYDROXYUNDECANOIC LACTONE and MACROCYCLIC
POLYAMINES: 1,4,7,10,13,16-HEXAAZACYCLOOCTADE-
CANE.

Nitrogen heterocycles continue to be valuable reagents and
provide new synthetic approaches such as NITRONES FOR
INTRAMOLECULAR-1,3-DIPOLAR CYCLOADDITIONS:
HEXAHYDRO-1,3,3,6-TETRAMETHYL-2,1-BENZISOXAZO-
LINE. Substituting on a pyrrolidine can be accomplished by
using NUCLEOPHILIC α-sec-AMINOALKYLATION: 2-(DI-
PHENYLHYDROXYMETHYL)PYRROLIDINE. Arene oxides
have considerable importance for cancer studies, and the example
ARENE OXIDE SYNTHESIS: PHENANTHRENE 9,10-OXIDE
has been included. An aromatic reaction illustrates RADICAL
ANION ARYLATION: DIETHYL PHENYLPHOSPHONATE.

Bicyclic and cyclic ketones are also useful intermediates and a
series of preparations are given in BICYCLIC KETONES FOR
TROPINONE SYNTHESIS: 2α,4α-DIMETHYL-8-OXABI-
CYCLO[3.2.1]OCT-6-EN-3-ONE; CYCLOPENTENONES
FROM α,α'-DIBROMOKETONES AND ENAMINES: 2,5-DI-
METHYL-3-PHENYL-2-CYCLOPENTEN-1-ONE; γ-KETO-
ESTER TO PREPARE CYCLIC DIKETONES: 2-METHYL-
1,3-CYCLOPENTANEDIONE, with the companion preparation
γ-KETOESTERS FROM ALDEHYDES VIA DIETHYL
ARYLSUCCINATES: 4-OXOHEXANOIC ACID ETHYL
ESTER, which gives the procedure used to prepare the needed

intermediate. OXIDATION OF ALCOHOLS BY METHYL SULFIDE-*N*-CHLOROSUCCINIMIDE-TRIETHYLAMINE: 4-*tert*-BUTYLCYCLOHEXANONE also demonstrates a general method for oxidizing alcohols to ketones. The use of boranes to prepare cyclic ketones has already been discussed. The trichloroacetyl substitution has been used in the procedure ALLYLICALLY TRANSPOSED AMINES FROM ALLYLIC ALCOHOLS: 3,7-DIMETHYL-1,6-OCTADIEN-3-AMINE. Alkylation of acetylenes alpha to the acetylene group is shown in the procedure 3-ALKYL-1-ALKYNES SYNTHESIS: 3-ETHYL-1-HEXYNE.

The Board of Editors welcomes both the submission of preparations for future volumes and suggestions for change that will enhance the usefulness of *Organic Syntheses*. Submitters are kindly asked to examine the instructions on pages v and vi that describe the type of preparations we wish to receive and also the information to be included in each contribution. A style guide for preparing manuscripts is available from the Secretary to the Board, and submitters are requested to follow its instructions.

As in previous volumes of *Organic Syntheses* unchecked procedures are tabulated at the end of this volume. Of the preparations received between July 1, 1977, and June 30, 1978, only those that have been accepted by the Board of Editors for checking are listed. These unchecked procedures are available from the Secretary's office for a nominal fee.

The Editor-in-Chief wishes to acknowledge a number of people for their efforts on behalf of *Organic Syntheses*. First, I would like to acknowledge the submitters who have generously agreed to openly share their experimental expertise in the precise manner required by *Organic Syntheses* and who have patiently borne with us during the checking and editing process. My colleagues on the Board of Editors and their collaborators have checked many of the procedures included in this volume. I would like to take this opportunity to warmly thank my own co-workers for sharing with me a belief in the importance of this work and for contributing their laboratory efforts to this and other volumes of *Organic Syntheses*.

Professor Wayland Noland, Secretary to the Board, has continued to provide the extensive coordination required in the preparation of a volume such as this. Special thanks must be given to Mrs. Rita Seelig Ayers, who, as an expert information chemist, put in long hours checking *Chemical Abstracts* names and providing very valuable assistance in the final proofing of the manuscript. Finally, but not least, I want to thank my very able secretary, Mrs. Susan S. Wilson, who, with the help of Miss Elizabeth Ann Knipmeyer and Mrs. Mary Wooten, typed the manuscript.

WILLIAM A. SHEPPARD

Wilmington, Delaware
March 1978

CONTENTS

ORGANIC SYNTHESES

3-ALKYL-1-ALKYNES SYNTHESIS:
3-ETHYL-1-HEXYNE

$$C_4H_9C\equiv CH \xrightarrow[\text{2. } C_2H_5Br, \text{ 0-25}^\circ]{\text{1. } C_4H_9Li, \text{ pentane}}$$

Submitted by A. J. QUILLINAN and F. SCHEINMANN[1]
Checked by Y. KITA and G. BÜCHI

1. Procedure

A dry 2-l., three-necked, round-bottomed flask is fitted with a nitrogen inlet, a reflux condenser provided with a gas outlet connected to a gas–bubbler, a rubber septum, and a magnetic stirrer. After being charged with 400 ml. of pure, dry pentane (Note 1) and 41 g. (0.50 mole) of 1-hexyne (Note 1), the flask is flushed with nitrogen, immersed in a cold bath (Note 2), and the contents stirred. The nitrogen atmosphere is maintained, and a solution of butyllithium in hexane or pentane (500 ml. of a $2.5N$ solution, or 1.25 mole) is transferred to the flask with a 100-ml. syringe or a cannula (Note 3). The mixture is allowed to warm to 10° and stirred for 30 minutes, until the initially formed precipitate has completely dissolved.

The clear yellow solution is recooled to 0° in an ice bath, and 88 g. (0.80 mole) of freshly–distilled ethyl bromide (Note 4) in 100 ml. of pure pentane is added dropwise with stirring over a 30-minute period while the solution warms to room temperature. Formation of a precipitate begins after an hour and is virtually complete after 6 hours (Note 5). After the mixture has been stirred for 2 days, 400 ml. of $4N$ hydrochloric acid is carefully added with cooling (ice bath) and stirring. The layers are separated, and the organic phase (Note 6) is washed with 15 ml. of water, dried over anhydrous potassium carbonate, and filtered. Low-boiling materials are removed by distillation through an efficient Vigreux column (Note 7). The residue is distilled using a spinning-band apparatus (Note 8), to give 35.2–35.8 g. (64–65%) of 3-ethyl-1-hexyne as a colorless, pungent oil (Note 9), b.p. 113–114°, n^{20}D 1.4101 (Note 10).

2. Notes

1. Solvents must be pure since higher-boiling impurities will accumulate in the reaction product, making the final distillation much more difficult and reducing the purity of the product. Commercial pentane was purified by redistillation through an efficient Vigreux column (Note 7) and collected at 33–34°. 1-Hexyne is available from Fluka A G, Tridom Chemical, Inc., and Koch-Light Laboratories, Ltd., or by synthesis from sodium acetylide and 1-bromobutane (L. Brandsma, "Preparative Acetylene Chemistry," Elsevier, Amsterdam, 1971, p. 45.).

2. The temperature is not critical; the submitters used a bath temperature of −35 to −20°, but claimed that an acetone–dry ice bath is satisfactory or that an ice-water bath with slow addition of butyllithium can also be used.

3. All operations involving alkyllithium reagents should be carried out under an inert gas, since they tend to ignite spontaneously in air.

4. Commercial ethyl bromide was dried over anhydrous potassium carbonate, filtered, and redistilled prior to use.

5. The reaction may be conveniently terminated here, but reaction for two days gives a slightly higher yield at the expense of some hydrocarbon impurity formed from the slow reaction of excess alkyllithium with the alkyl bromide. If this impurity can contaminate the final product, the side reaction may be suppressed by conducting the reaction in the minimum amount of solvent necessary to dissolve the dilithium complex.

6. Dark coloration is usually a result of insufficient acidity; the aqueous phase at the separation stage should have a pH between 2 and 4.

7. A vacuum-jacketed Vigreux column of 1.5-cm. internal diameter and ca. 90-cm. length is satisfactory. With shorter or less efficient columns, redistillation of the distillate may be necessary to reduce losses in yield.

8. The submitters used a Büchi spinning-band distillation apparatus by Abegg of approximately 30 theoretical plates resolution, operated at a reflux ratio of 10–15:1 and having a capacity of 50–100 ml. The checkers found that this distillation required four days in order to obtain the pure product.

9. 3-Ethyl-1-hexyne is an irritant to the membranes of the nose

and throat and should only be handled in a well-ventilated hood. Other 3-substituted 1-alkynes prepared in this series[2] have pleasant or neutral odors, but direct contact should be avoided since toxicity data are not available.

10. The previously unknown 3-ethyl-1-hexyne was further characterized by the submitters as follows: infrared cm.$^{-1}$: 3338, 2975, 2940, 2881, 2103, 1460, 1380, 1240; ^{13}C magnetic resonance δ (chloroform-d, downfield from tetramethylsilane): 88.0, 69.23, 37.12, 33.29, 28.23, 20.69, 13.97, 11.71; proton magnetic resonance (chloroform-d) δ (multiplicity, numbers of protons, assignment, coupling constant J in Hz.): 0.99 (triplet, 6, CH_3, $J = 6$), 1.2–1.7 (multiplet, 6, CH_2), 2.00 (doublet, 1, $C{\equiv}CH$, $J = 2$); 2.3 (multiplet, 1, tertiary CH).

Hydrogenation at 20° over Adams catalyst, platinum oxide, at atmospheric pressure afforded the known 3-ethylhexane,[3] b.p. 118–119° (literature[2] b.p. 119°).

In an analogous synthesis,[2a] 3-butyl-1-heptyne was also characterized by mercuric oxide–sulfuric acid hydration[4] to the known 3-butyl-2-heptanone,[5,6] which also formed the known semicarbazone.[6]

3. Discussion

This procedure has been shown[2] to be extremely general and applicable to a wide variety of straight-chain 1-acetylenes, 4-substituted 1-acetylenes, and α,ω-diacetylenes, together with primary halides, sterically hindered primary halides, secondary halides, and α,ω-dihalides.

Most of the compounds formed are new and were formerly inaccessible,[7] being only available by dehydrohalogenation of geminal or 1,2-dibromides which are often unavailable themselves.[7] Alcoholic potassium hydroxide[8,9] or sodamide in liquid paraffin[7,10] under forceful conditions has been used for this elimination, but yields are generally not good.[7–10]

The procedure described here is characterized by good yields, mild conditions, and easy synthesis of a pure form from readily available starting materials. Since tertiary aliphatic acetylenes do not form readily under these conditions, the excess of alkyllithium used is not particularly critical. The small amount of by-products that also form is similarly readily removed at the distillation stage.

1. Department of Chemistry and Applied Chemistry, University of Salford, Salford M5 4WT, England.
2. (a) A. J. Quillinan, E. A. Khan, and F. Scheinmann, *Chem. Commun.*, 1030 (1974); (b) J. Klein and S. Brenner, *J. Org. Chem.*, **36,** 1319 (1971); (c) J. Klein and J. Y. Becker, *Tetrahedron,* **28,** 5385 (1972); (d) J. Klein and J. Y. Becker, *J. Chem. Soc., Perkin Trans. II,* 599 (1973).
3. L. Clarke and E. R. Riegel, *J. Amer. Chem. Soc.,* **34,** 674 (1912).
4. R. M. Roberts, J. C. Gilbert, L. B. Rodewald, and A. S. Wingrove, "An Introduction to Modern Experimental Organic Chemistry", Holt, Rinehart and Winston, New York, N. Y., 1969.
5. K. Hess and R. Bappert, *Justus Liebigs Ann. Chem.,* **441,** 151 (1925).
6. W. B. Renfrow, Jr., *J. Amer. Chem. Soc.,* **66,** 144 (1944).
7. T. L. Jacobs, *Org. React.* **5,** 1 (1949).
8. V. Sawitsch, *C. R. H. Seances Acad. Sci.,* **52,** 399 (1861).
9. W. Morkownikoff, *Bull. Soc. Chim. Fr.,* **14,** 90 (1861).
10. B. Gredy, *Bull. Soc. Chim. Fr.* [5], **2,** 1951 (1935).

Appendix

Chemical Abstracts Nomenclature (Collective Index Number; Registry Numbers)

1-Hexyne, 3-ethyl- (8,9); (−)

1-Hexyne (8,9); (693-02-7)

Lithium, butyl- (8,9); (109-72-8)

Ethyl bromide: Ethane, bromo- (8,9); (74-96-4)

Hexane, 3-ethyl- (8,9); (619-99-8)

1-Heptyne, 3-butyl- (8,9); (−)

2-Heptanone, 3-butyl- (8,9); (997-69-3)

ALLYLICALLY TRANSPOSED AMINES FROM ALLYLIC ALCOHOLS: 3,7-DIMETHYL-1,6-OCTADIEN-3-AMINE

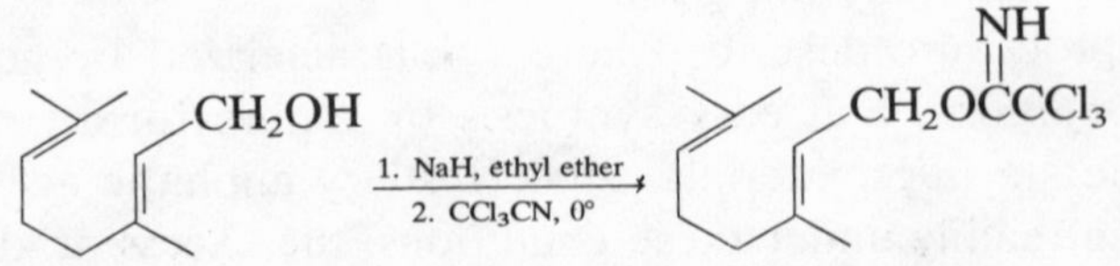

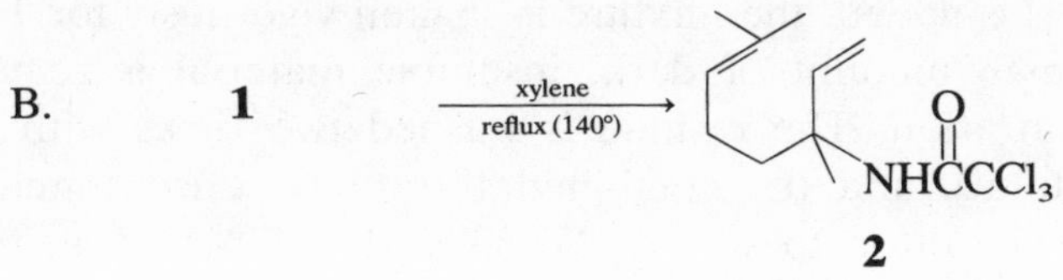

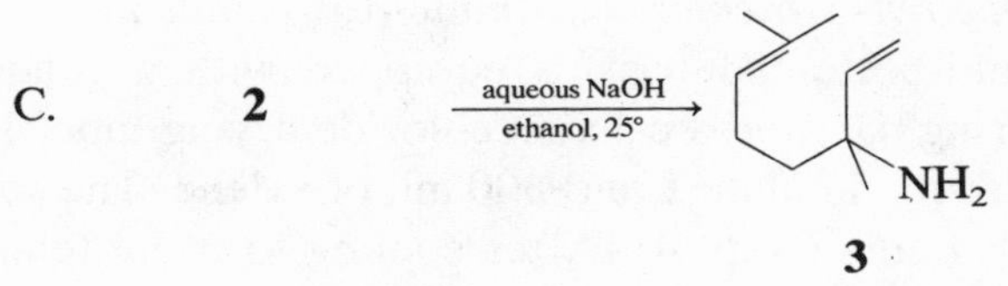

B.　　1　$\xrightarrow[\text{reflux (140°)}]{\text{xylene}}$

2

C.　　2　$\xrightarrow[\text{ethanol, 25°}]{\text{aqueous NaOH}}$

3

Submitted by LANE A. CLIZBE and LARRY E. OVERMAN[1]
Checked by A. BROSSI, H. MAYER, and N. KAPPELER

1. Procedure

Caution! Part A should be carried out in a well-ventilated hood to avoid exposure to trichloroacetonitrile vapors.

A. *Geraniol trichloroacetimidate* (**1**). A dry 250-ml., three-necked flask is equipped with a magnetic stirring bar, a pressure-equalizing dropping funnel, a thermometer, and a nitrogen inlet tube. The apparatus is flushed with nitrogen and charged with 410 mg. (0.010 mole) of sodium hydride dispersed in mineral oil (Note 1) and with 15 ml. of hexane. The suspension is stirred, and the hydride is allowed to settle. The hexane is removed with a long dropping pipette, and 60 ml. of anhydrous ethyl ether is added. A solution of 15.4 g. (0.10 mole) of geraniol and 15 ml. of anhydrous ethyl ether is added over 5 minutes. After the evolution of hydrogen ceases (less than 5 minutes), the reaction mixture is stirred for an additional 15 minutes. The clear solution is then cooled to between −10 and 0° in an ice–salt bath. Trichloro-acetonitrile (10.0 ml., 14.4 g., 0.10 mole) is added dropwise to the stirred solution, while the reaction temperature is maintained below 0° (Note 2). Addition is completed within 15 minutes, and the reaction mixture is allowed to warm to room temperature. The light amber reaction mixture is poured into a 250-ml., round-bottomed flask, and the ethyl ether is removed with a rotary evaporator. Pentane [150 ml., containing 0.4 ml. (0.010 mole) of methanol] is added, the mixture is shaken vigorously for 1 minute, and a small amount of dark, insoluble material is removed by gravity filtration. The residue is washed two times with pentane (50 ml. total), and the combined filtrate is concentrated with a rotary evaporator to afford 27–29 g. (90–97%) of nearly pure (Note 3) imidate **1**.

B. 3,7-*Dimethyl-3-trichloroacetamido*-1,6-*octadiene*   (**2**). A 500-ml., round-bottomed flask is equipped with a condenser, a magnetic stirring bar, and a calcium chloride drying tube. The flask is charged with the imidate **1** and 300 ml. of xylene. The solution is refluxed for 8 hours (Note 4). After cooling to room temperature the dark xylene solution is filtered through a short column (4.5 cm. in diameter) packed with silica gel (70 g.) and toluene. The column is eluted with an additional 250 ml. of toluene, and the combined light yellow eluant is concentrated with a rotary evaporator. Vacuum distillation through a 15-cm. Vigreux column yields 20–22 g. (67–74% for the two steps) of the octadiene **2** as a colorless liquid, b.p. 94–97° (0.03 mm.) (Note 5).

C. 3,7-*Dimethyl*-1,6-*octadien-3-amine* (**3**).   A 500-ml., round-bottomed flask is equipped with a magnetic stirring bar, a condenser, and a nitrogen inlet tube. The flask is charged with 9.0 g. (0.030 mole) of the octadiene **2**, 160 ml. of 95% ethanol, and 150 ml. of aqueous 6N sodium hydroxide solution. The air is replaced with nitrogen (Note 6), and the solution is stirred at room temperature for 40 hours (Note 7). Ethyl ether (300 ml.) is added, the organic layer is separated, and the aqueous layer is washed twice with 50 ml. of ethyl ether. After drying over anhydrous potassium carbonate and filtration, the organic extracts are concentrated with a rotary evaporator to afford a white, semisolid residue. This residue is extracted four times with 50 ml. of boiling hexane. The extract is concentrated with a rotary evaporator, and the residual yellow liquid is distilled under reduced pressure using a short-path apparatus to yield 2.99–3.45 g. (65–75%) of the amine **3**, b.p. 58–61° (2.6 mm.) (Notes 8 and 9).

2. Notes

1. The reagents used in this procedure were obtained from the following sources: geraniol (99+%), Aldrich Chemical Company,

Inc.; sodium hydride (58% dispersion in mineral oil), Alfa Products, Division of the Ventron Corporation; trichloroacetonitrile, Aldrich Chemical Company, Inc.; xylene (a mixture of isomers, b.p. 137–144°), Mallinckrodt Chemical Works; anhydrous ethyl ether, Mallinckrodt Chemical Works; Silica Gel (Grade 62), Grace Davidson Chemical. The reagents used by the checkers were obtained from the following sources: geraniol (98.7%), sodium hydride (58% dispersion in mineral oil), trichloroacetonitrile, and xylene (a mixture of isomers, b.p. 137–143°), Fluka A G, Chemische Fabrik, Buchs, Switzerland; Silica gel (70–230 mesh ASTM), E. Merck A G, Darmstadt, Germany.

2. For secondary or tertiary alcohols, the yields are improved by adding the alcohol–alkoxide solution dropwise to a solution of trichloroacetonitrile and ethyl ether at 0°.

3. The crude imidate **1** is sufficiently pure (checked by proton magnetic resonance) for most purposes and can be used in the next step without further purification. Infrared (neat) cm.$^{-1}$: 3340 weak (N—H), 1660 strong (C=N); proton magnetic resonance (carbon tetrachloride) δ (multiplicity, number of protons, assignment, coupling constant J in Hz.): 8.3 (broad singlet, 1.0, NH), 5.5 (approximate triplet, 1.0, H_2, $J = 7$), 5.1 (broad singlet, 1.0, H_6), 4.77 (doublet, 2.0, H_1, $J = 7$), 1.73 (singlet, 3.0, CH_3), 1.66 (singlet, 3.0, CH_3), 1.58 (singlet, 3.0, CH_3). The crude imidate **1** may be distilled rapidly through a short Vigreux column to give 24–28 g. (80–93%) of distilled product, b.p. 109–111° (0.1 mm.). However, the checkers found by proton magnetic resonance analysis that this product already contains substantial amounts of the octadiene **2** formed by thermal rearrangement during distillation.

4. The reaction can be conveniently monitored in the infrared by observing the decrease in the C=N stretching absorption at 1660 cm.$^{-1}$ or by thin-layer chromatography [silica, developed with hexane–ethyl acetate (9:1) or (4:1)].

5. The octadiene **2** appears pure by proton magnetic resonance and elemental analysis. A thin-layer chromatogram [silica, developed with hexane–ethyl acetate (4:1), visualized with 5% ceric ammonium nitrate in 20% sulfuric acid and heating] shows a major spot at $Rf = 0.4$ and a very small impurity at $Rf = 0.7$. The trace impurity may be removed by crystallization from hexane at −78°. The checkers found that a gas chromatogram (2% Silicone OV-17

on Gaschrom Q., 80–100 mesh, 200×0.22 cm., 120°) showed one major peak (97.0–97.2%) at a relative retention value of 1.00 (17.3 minutes, nitrogen, 30 ml. per minute), and a minor peak at 0.94 (<3%). The octadiene 2 has the following spectral properties: infrared (neat) cm.$^{-1}$: 3423 and 3355 weak (N—H), 1722 strong (C=O), 1504 strong (CO—NH—, amide II band), 983 weak and 915 medium (CH=CH$_2$); proton magnetic resonance (carbon tetrachloride) δ (multiplicity, number of protons, assignment, coupling constant J in Hz.): 6.6 (broad singlet, 1.0, NH), 5.96 (approximate doublet of doublets, 1.0, H_2, J = 9.5 and 18.5), 4.9–5.3 (multiplet, 3.0, H_1 and H_6), 1.67 (singlet, 3.0, CH_3), 1.60 (singlet, 3.0, CH_3), 1.49 (singlet, 3.0, CH_3); ^{13}C magnetic resonance (acetone-d_6, tetramethylsilane reference) δ (assignment): 159.8 (C=O), 141.1 (C_2), 132.4 (C_7), 123.4 (C_6), 113.2 (C_1), 93.3 (CCl$_3$), 58.7 (C_3), 38.8 (C_4), 25.7 (CH_3), 23.9 (CH_3), 22.5 (C_5), 17.5 (CH_3).

6. The apparatus described by W. S. Johnson and W. P. Schneider *Org. Syn.*, Coll. Vol. **4**, 132 (1963) was used to maintain a nitrogen atmosphere.

7. The reaction time may be cut to less than 1 hour by running the reaction at reflux, but the yield is 5–10% lower. The checkers found that under these conditions the weight yield of the distilled product was 67–70%. However a gas chromatogram (3% Silicone GE-SE-30 on Gaschrom Q, 80–100 mesh, 200×0.22 cm., 60°) showed that the product contained substantial amounts (19–23%) of an unidentified by-product having a relative retention value of 1.22 (nitrogen, 30 ml. per minute) and a minor peak at 0.96 (<2%). The checkers also found that the reaction can be conveniently monitored by thin-layer chromatography [silica, hexane-ethyl acetate (4:1)].

8. The checkers found that the crude amine 3 (linalylamine) is more conveniently distilled at a pressure of 11–12 mm.

9. A gas chromatogram (10% Carbowax 20 M–2% KOH on Chromosorb W, AW, 80–100 mesh, 180×0.3 cm., 90°) showed one major peak (99%) at a relative retention value of 1.00 (3.5 minutes, nitrogen, 50 ml. per minute), and a minor peak at 1.1 (<1%). Temperature programming to 200° detected the presence of several higher-boiling trace impurities (<1%). The checkers found that a gas chromatogram (3% Silicone GE-SE-30 on Gas-

chrom Q, 80–100 mesh, 200×0.22 cm., 60°) showed one major peak (97.9–98.4%) at a relative retention value of 1.00 (11.8 minutes, nitrogen, 30 ml. per minute), and a minor peak at 0.96 (<2.1%). Infrared (neat) cm.$^{-1}$: 3345 and 3301 weak (N—H), 996 and 912 medium (CH=CH$_2$); proton magnetic resonance (carbon tetrachloride) δ (multiplicity, number of protons, assignment, coupling constant J in Hz.): 5.84 (approximate doublet of doublets, 1.0, H_2, $J = 10.1$ and 17.8), 4.7–5.2 (multiplet, 3.0, H_1 and H_6), 1.63 (singlet, 3.0, CH$_3$), 1.55 (singlet, 3.0, CH$_3$), 1.08 (singlet, 3.0, CH$_3$).

3. Discussion

This procedure illustrates a general method for the preparation of rearranged allylic amines from allylic alcohols.[2,3] The method is experimentally simple and has been used to prepare a variety of allylic *prim-*, *sec-*, and *tert*-carbonyl amines as illustrated in Table I. The only limitation encountered so far is a competing ionic elimination reaction which becomes important for trichloroacetimidic esters of 3-substituted-2-cyclohexen-1-ols.[3,4] The rearrangement is formulated as a concerted [3,3]-sigmatropic rearrangement on the basis of its stereo- and regiospecificity[3,5] which are

TABLE I

CONVERSION OF ALLYLIC ALCOHOLS INTO REARRANGED TRICHLORO-ACETAMIDES[3]

Alcohol	Trichloroacetamide Product	Overall Isolated Yield (%)
(E)-2-Hexen-1-ol	3-Trichloroacetamido-1-hexene	72
Cinnamyl alcohol	3-Phenyl-3-trichloroacetamido-1-propene	76
1-Hepten-3-ol	(E)-1-Trichloroacetamido-2-heptene	74
2-Cyclohexen-1-ol	3-Trichloroacetamido-1-cyclohexene	61
Linalool	3,7-Dimethyl-1-trichloroacetamido-2,6-octadiene^a	83
3,5,5-Trimethyl-2-cyclohexen-1-ol	3,5,5-Trimethyl-3-trichloroacetamido-1-cyclohexene	10–43
3-(4,4-Ethylenedioxybutyl)-2-cyclohexen-1-ol	3-(4,4-Ethylenedioxybutyl)-3-trichloroacetamido-1-cyclohexene	20

^{a} A mixture of E and Z isomers.

similar to those observed in related [3,3]-sigmatropic processes.[6] In certain cases the allylic imidate rearrangement may be accomplished at or below room temperature by the addition of catalytic amounts of mercuric salts.[2,3]

The experimental procedure for the addition of an alcohol to trichloroacetonitrile, a modification of the procedure of Cramer,[7] appears to be totally general. For hindered secondary or tertiary alcohols it is more convenient to form the catalytic alkoxide with potassium hydride, and it is essential that the alcohol–alkoxide solution be added to (inverse addition) a solution of trichloro-acetonitrile in ethyl ether at 0°.[3] The thermal rearrangement of allylic imidates was first reported by Mumm and Möller[8] in 1937, and has been noted in scattered reports since that time.[9] In these cases the imidates were not available by a general route (as is the case for trichloroacetimidates), and as a result this rearrangement had not become a generally useful synthetic method. Alternative methods for the allylic transposition of oxygen and nitrogen functions include the thermolysis of allylic sulfamate esters,[10] phenylurethanes,[11] oxime O-allyl ethers,[12] and N-chloro-sulfonylurethanes,[13] as well as the S_N2' reaction of certain allylic alcohol esters with amines.[14] These procedures are much less attractive than the method reported here, usually affording mixtures of allylic isomers.[10-14] Other routes for the preparation of 3,7-dimethyl-1,6-octadien-3-amine have not, to our knowledge, been reported.

1. Department of Chemistry, University of California, Irvine, California, 92717.
2. L. E. Overman, J. Amer. Chem. Soc., **96**, 597 (1974).
3. L. E. Overman, J. Amer. Chem. Soc., **98**, 2901 (1976).
4. L. E. Overman, Tetrahedron Lett., 1149 (1975).
5. Y. Yamamoto, H. Shimoda, J. Oda, and Y. Inouye, Bull. Chem. Soc. Jap., **49**, 3247 (1976).
6. S. J. Rhoads and N. R. Raulens, Org. React., **22**, 1 (1975).
7. F. Cramer, K. Pawelzik, and H. J. Baldauf, Chem. Ber., **91**, 1049 (1958).
8. O. Mumm and F. Möller, Chem. Ber., **70**, 2214 (1937).
9. W. M. Lauer and R. G. Lockwood, J. Amer. Chem. Soc., **76**, 3974 (1954); W. M. Lauer and C. S. Benton, J. Org. Chem., **24**, 804 (1959); R. M. Roberts and F. A. Hussein, J. Amer. Chem. Soc., **82**, 1950 (1960); D. St. C. Black, F. W. Eastwood, R. Okraglik, A. J. Poynton, A. M. Wade, and C. H. Welker, Aust. J. Chem., **25**, 1483 (1972).
10. E. H. White and C. A. Elliger, J. Amer. Chem. Soc., **87**, 5261 (1965).
11. M. E. Synerholm, N. W. Gilman, J. W. Morgan, and R. K. Hill, J. Org. Chem., **33**, 1111 (1968).
12. A. Eckersley and N. A. J. Rogers, Tetrahedron Lett., 1661 (1974).

ALLYLICALLY TRANSPOSED AMINES FROM ALLYLIC ALCOHOLS   11

13. J. B. Hendrickson and I. Joffee, *J. Amer. Chem. Soc.*, **95**, 4083 (1973).
14. G. Stork and W. N. White, *J. Amer. Chem. Soc.*, **78**, 4609 (1965); G. Stork and A. F. Kreft, III, *J. Amer. Chem. Soc.*, **99**, 3850 (1977).

Appendix
Chemical Abstracts Nomenclature (Collective Index Number; Registry Numbers)

1,6-Octadien-3-amine, 3,7-dimethyl- (**3**) (8,9); (59875-02-4)

Geraniol Trichloroacetimidate (**1**); (−)

Sodium hydride (8,9); (7646-69-7)

Geraniol: 2,6-Octadien-1-ol, 3,7-dimethyl-, (*E*)- (8,9); (106-24-1)

Trichloroacetonitrile: Acetonitrile, trichloro- (8,9); (545-06-2)

3,7-Dimethyl-3-trichloroacetamido-1,6-octadiene (**2**): Acetamide, 2,2,2-trichloro-*N*-(3,7-dimethyl-1,6-octadien-3-yl)- (8,9); (−)

Xylene (8); Benzene, dimethyl- (9); (1330-20-7)

Toluene (8); Benzene, methyl- (9); (108-88-3)

Potassium carbonate: Carbonic acid, dipotassium salt (8, 9); (584-08-7)

Potassium hydride (8,9); (7693-26-7)

3-Trichloroacetamido-1-hexene: Acetamide, 2,2,2-trichloro-*N*-(1-hexen-3-yl)- (8,9); (−)

3-Phenyl-3-trichloroacetamido-1-propene: Acetamide, 2,2,2-trichloro-*N*-(1-phenyl-2-propenyl)- (8, 9); (59874-90-7)

(*E*)-1-Trichloroacetamido-2-heptene: Acetamide, 2,2,2-trichloro-*N*-(2-hepten-1-yl)-, (*E*); (8,9); (−)

3-Trichloroacetamido-1-cyclohexene: Acetamide, 2,2,2-trichloro-*N*-(2-cyclohexen-1-yl)- (8,9); (−)

3,7-Dimethyl-1-trichloroacetamido-2,6-octadiene: Acetamide, 2,2,2-trichloro-*N*-(3,7-dimethyl-2,6-octadienyl)- (8,9); (*E*) (59874-96-3) (*Z*) (59874-97-4)

3,5,5-Trimethyl-3-trichloroacetamide-1-cyclohexene: Acetamide, 2,2,2-trichloro-*N*-(1,5,5-trimethyl-2-cyclohexen-1-yl)- (8,9); (59874-95-2)

3-(4,4-Ethylenedioxybutyl)-3-trichloroacetamido-1-cyclohexene: (−)

ARENE OXIDE SYNTHESIS: PHENANTHRENE 9, 10-OXIDE

(Phenanthro[9,10-b]oxirene, 1a,9b-dihydro)

A.

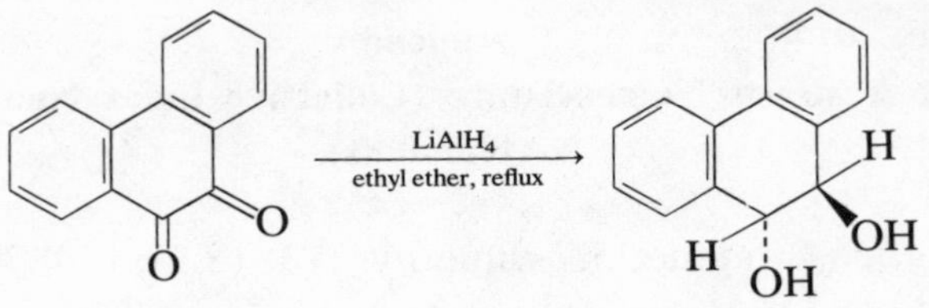

B.   1

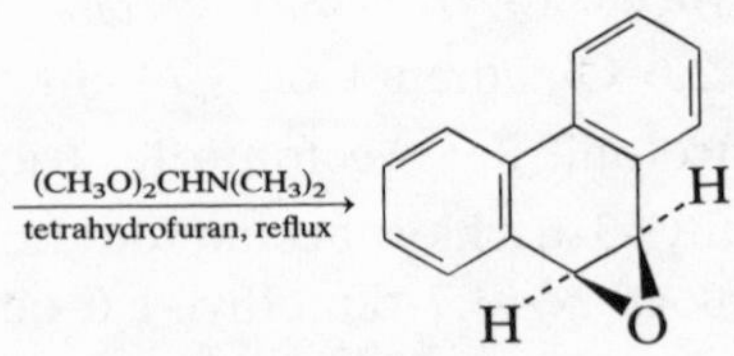

Submitted by CECILIA CORTEZ and RONALD G. HARVEY[1]
Checked by JAMES JACKSON and ORVILLE L. CHAPMAN

1. Procedure

Caution! See benzene warning, p. 168.

A. 9,10-*Dihydro*-trans-9,10-*phenanthrenediol*   (**1**). Phenan-threnequinone (6 g., 0.029 mole) (Note 1) is placed in a fritted-glass (coarse porosity) extraction thimble of a Soxhlet apparatus over a 1-l. flask containing a suspension of 3 g. of lithium aluminum hydride in 500 ml. of anhydrous ethyl ether (Note 2). Extraction of the quinone over a period of 16 hours affords a green solution (Note 3). The reaction is quenched by the cautious addition of water (Note 4) and neutralized by the addition of glacial acetic acid. The ether layer is separated, and the aqueous layer is extracted with two 200-ml. portions of ethyl ether. The combined ether extracts are washed consecutively with aqueous sodium bicarbonate and water, and then dried over magnesium sulfate. Evaporation of the solvent under reduced pressure gives the crude product (Note 5), which is recrystallized from benzene to give essentially pure diol **1** as fluffy white needles, m.p. 185–190°, in a yield of 3.8–4.1 g. (62–68%) (Note 6).

B. *Phenanthrene*-9,10-*oxide* (**2**). A solution of 10.6 g. (0.05 mole) of **1** and 13 g. of N,N-dimethylformamide dimethyl acetal (Note 7) in 40 ml. of N,N-dimethylformamide (Note 8) and 100 ml. of dry tetrahydrofuran (Note 9) is heated at reflux for 16 hours. The solution is then allowed to cool to room temperature, and 200 ml. of water and 100 ml. of ethyl ether are added. The organic layer is then separated, the aqueous layer is washed with two 200-ml. portions of ethyl ether, and the combined ether phases are dried over magnesium sulfate. Evaporation of the solvent under reduced pressure gives 9.6 g of a yellow solid. Trituration with 25 ml. of hexane removes colored impurities. Recrystallization from benzene–cyclohexane (Note 10) gives the oxide **2** as off-white plates, m.p. 125° (dec.) (Note 11), 5.6–6.2 g. (58–64%). A second crop of 1.0 g can be obtained, for an overall yield of 68–74% (Note 12).

2. Notes

1. Phenanthrenequinone free of anthraquinone is available from Aldrich Chemical Company, Inc., or from J. T. Baker Chemical Company. It should be recrystallized from benzene before use.

2. Use of more efficient solvents (tetrahydrofuran, isopropyl ether, dimethoxyethane) or more soluble metal hydride reagents (sodium borohydride, lithium tributoxy aluminum hydride, sodium bis(2-methoxyethyl) aluminum hydride) favors the alternative reduction pathway to the hydroquinone.

3. The checkers noted that use of a paper thimble resulted in increased time for extraction. The submitters recommend use of a glass thimble, since prolonged heating can lead to lower yields. It is easier to determine when extraction is complete with a transparent thimble. Other quinones may require longer extraction periods.

4. Care must be taken to add the water cautiously and slowly, since the reaction between water and lithium aluminum hydride is vigorous. The reaction is quenched when the solution stops refluxing.

5. The crude product may darken on drying because of the presence of minor amounts of the air-sensitive hydroquinone by-product.

6. Lower yields usually result in large-scale reactions. The checkers obtained product of m.p. 189–191° in runs with slightly lower yields.

7. *N,N*-Dimethylformamide dimethyl acetal, obtained from Aldrich Chemical Company, Inc., is redistilled before use.

8. *N,N*-Dimethylformamide is distilled under reduced pressure and stored over molecular sieves, type 4Å.

9. Tetrahydrofuran is distilled from lithium aluminum hydride.

10. Excessive heating during recrystallization should be avoided because it can lead to thermal decomposition of the product.

11. Because of the relative facility of thermal rearrangement to phenols, melting points of arene oxides are not an entirely reliable index of purity. The checkers found variation from 119 to 135° (dec.). Purification by chromatography on activity IV alumina is also possible, but residence time on the column should be held to a minimum.

12. The proton magnetic resonance spectrum (chloroform-*d*) of the pure product **2** showed a characteristic oxiranyl singlet peak at δ 4.67 (singlet, 2*H*) and an aromatic signal at 7.2–7.8 (multiplet, 8*H*).

3. Discussion

The method employed here is essentially that reported earlier,[2] modified by subsequent experience.[3] In the second step, *N,N*-dimethylformamide dimethyl acetal acts as a dehydrating agent to convert the diol to epoxide, and is converted to *N,N*-dimethylformamide and methanol. Phenanthrene-9,10-oxide also has been prepared by cyclization of 2,2′-biphenyldicarboxaldehyde with hexamethylphosphorus triamide[4] and by dehydrohalogenation of the acetate of 10-chloro-9,10-dihydro-9-phenanthrenol obtained through reaction of the corresponding 2-alkoxy-1,3-dioxolane with trimethylsilyl chloride.[5] The present procedure is simpler, requiring fewer steps from readily available starting materials; both alternative procedures must start with phenanthrene. The product is relatively easy to purify since the only by-products are *N,N*-dimethylformamide and methanol (an important consideration with molecules sensitive to decomposition), and appears to be more stable on storage than the compound obtained via the dialdehyde route.

The cyclization method utilized in this synthesis appears quite general in its applicability, having been applied successfully in our laboratory[3] to the preparation of the K-region arene oxides[6] of benz[a]anthracene, chrysene, dibenz[a,h]anthracene, benzo[c]phenanthrene, pyrene, 1-methylphenanthrene, benzo-[a]pyrene, and 7,12-dimethylbenz[a]anthracene, among others. The latter two are potent carcinogens; the K-region oxides of these have been shown to be formed metabolically and to exhibit significant biological activity.[7]

The K-region quinones required as starting materials in this synthesis are in certain cases (e.g., phenanthrene, chrysene, benzo[c]phenanthrene) available from direct oxidation of the parent hydrocarbons with chromic acid. When oxidation occurs preferentially elsewhere in the molecule, the K-region dihydroaromatic derivatives can often be converted to the corresponding quinone through oxidation with dichromate in acetic acid–acetic anhydride[8], yields, however, are only in the 20–30% range. Alternatively, the K-region quinones may be obtained from the hydrocarbons through oxidation with osmium tetroxide to the corresponding cis-diols, followed by a second oxidation with pyridine–sulfur trioxide and dimethyl sulfoxide.[2,3] This is the most generally useful procedure. A significant advantage is that all possible K-region oxidized derivatives (cis-diols, quinones, trans-diols, phenols,[3] and hydroquinones[9]) with the ring system intact can be obtained directly or by appropriate modification of the general sequence. The disadvantage of this method, and of any alternative procedure,[4,5] involving the cis-diol is that the hazardous and expensive reagent osmium tetroxide is employed.

1. Ben May Laboratory, University of Chicago, Chicago, Illinois 60637.
2. S. H. Goh and R. G. Harvey, J. Amer. Chem. Soc., 95, 242 (1973).
3. R. G. Harvey, S. H. Goh, and C. Cortez, J. Amer. Chem. Soc., 97, 3468 (1975).
4. M. S. Newman and S. Blum, J. Amer. Chem. Soc., 86, 5598 (1964).
5. P. Dansette and D. M. Jerina, J. Amer. Chem. Soc., 96, 1224 (1974).
6. The K-region of a polycyclic aromatic hydrocarbon is typified by the 9,10-bond of phenanthrene. According to the Schmidt–Pullman electronic theory, an unsubstituted K-region is a requirement for carcinogenic activity; see A. Pullman and B. Pullman, "La Cancerisation par les Substances Chimiques et la Structure Moleculaire," Masson, Paris, 1955.

7. For leading references see K. W. Jennette, A. M. Jeffrey, S. H. Blobstein, F. A. Beland, R. G. Harvey, and I. B. Weinstein, *Biochemistry*, **16,** 932 (1977).
8. H. Cho and R. G. Harvey, *Tetrahedron Lett.*, 1491 (1974).
9. H. Cho and R. G. Harvey, *J. Chem. Soc., Perkin Trans I*, 836 (1976).

Appendix
Chemical Abstracts Nomenclature; (Collective Index Number; Registry Numbers)

Phenanthrene-9,10-oxide (**2**): Phenanthrene, 9,10-epoxy-9,10-dihydro- (8); Phenanthro[9,10-*b*]oxirene, 1*a*,9*b*-dihydro- (9); (585-08-0)

9,10-Phenanthrenediol, 9,10-dihydro-, *trans-* (**1**) (8,9); (25061-61-4)

Phenanthrenequinone (8); 9,10-Phenanthrenedione (9); (84-11-7)

Lithium aluminum hydride: Aluminate (1-), tetrahydro, lithium (8), Aluminate (1-), tetrahydro-, lithium, (T-4)- (9); (16853-85-3)

N,N-Dimethylformamide dimethyl acetal: Trimethylamine, 1,1-dimethoxy- (8); Methanamine, 1,1-dimethoxy-*N,N*-dimethyl- (9); (4637-24-5)

Dimethylformamide: Formamide, *N,N*-dimethyl- (8,9); (68-12-2)

Furan, tetrahydro- (8, 9); (109-99-9)

2,2′-Biphenyldicarboxaldehyde (8); [1,1′-Biphenyl]-2,2′-dicarboxaldehyde (9); (1210-05-5)

Phosphorus triamide, hexamethyl- (8,9); (1608-26-0)

9-Phenanthrol, 10-chloro-9,10-dihydro-, acetate (8): 9-Phenanthrenol, 10-chloro-9,10-dihydro-, acetate (9); (1028-73-5)

Trimethylsilyl chloride: Silane, chlorotrimethyl- (8,9); (75-77-4)

Benz[*a*]anthracene (8,9); (56-55-3)

Chrysene (8,9); (218-01-9)

Dibenz[*a,h*]anthracene (8,9); (53-70-3)

Benzo[*c*]phenanthrene (8,9); (195-19-7)

Pyrene (8, 9); (129-00-0)

Phenanthrene, 1-methyl- (8,9); (832-69-9)

Benzo[*a*]pyrene (8,9); (50-32-8)

Benz[*a*]anthracene, 7,12-dimethyl- (8,9); (57-97-6)

Osmium tetroxide: Osmium oxide (OsO_4) (8,9); (20816-12-0)

Pyridine–sulfur trioxide: Pyridine, compound with sulfur trioxide (8), (28322-92-1); Pyridine, compound with sulfur trioxide (1:1) (9); (26412-87-3)

Dimethyl sulfoxide: Methyl sulfoxide (8); Methane, sulfinylbis- (9); (67-68-5)

BICYCLIC KETONES FOR TROPINONE SYNTHESIS: 2α,4α-DIMETHYL-8-OXABICYCLO[3.2.1]OCT-6-EN-3-ONE

A. $\underset{\text{O}}{\overset{\text{O}}{\text{CH}_3\text{CH}_2\overset{\|}{\text{C}}\text{CH}_2\text{CH}_3}} + 2\text{Br}_2 \xrightarrow[-10-0°]{\text{PBr}_3\text{ (catalyst)}} \text{CH}_3\text{CHBr}\overset{\|}{\text{C}}\text{CHBrCH}_3$
                                                                              1

B. 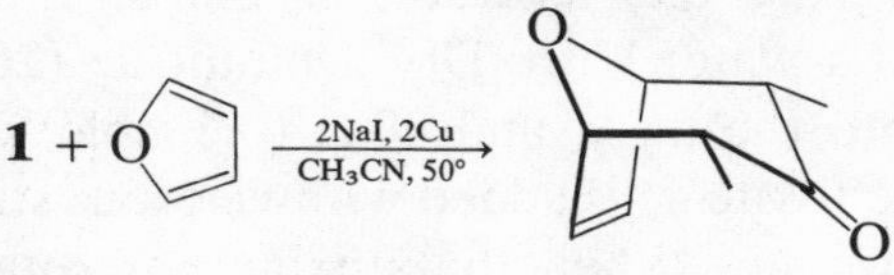 $\textbf{1} + $ $\xrightarrow[\text{CH}_3\text{CN, 50°}]{\text{2NaI, 2Cu}}$ $+ 2\text{HBr}$

2

Submitted by M. R. Ashcroft and H. M. R. Hoffmann[1]
Checked by D. M. Lokensgard and O. L. Chapman

1. Procedure

Caution! This reaction should be carried out in an efficient hood. 2,4-Dibromo-3-pentanone is a potent lachrymator and a readily absorbed skin irritant. Contact with the skin produces a sensation of sunburn and should be treated immediately by washing with a soap solution followed by washing with sodium bicarbonate solution.

A. *2,4-Dibromo-3-pentanone* (**1**). A three-necked, 250-ml. flask is fitted with a stirrer, a dropping funnel, and a condenser protected by a calcium chloride drying tube. Bromine (160 g., 1 mole) is added rapidly to a stirred solution of 45 g. (0.52 mole) of 3-pentanone (Note 1) and 1 ml. of phosphorus tribromide maintained between −10° and 0° with a dry ice–acetone bath in an efficient hood. Toward the end of the reaction very large amounts of hydrogen bromide are evolved, and the rate of addition must be controlled to allow the hood to exhaust the gas. Alternatively, a gas trap may be used. Depending on the efficiency of the hood, the addition should take 20–40 minutes. The flask is then evacuated with a water pump to remove dissolved hydrogen bromide, and the reaction mixture is immediately fractionally distilled through a 40-cm. column packed with glass helices (or, more quickly, with a Dufton column)[2] under reduced pressure. The dibromoketone 1, which is a mixture of *dl*- and *meso*-isomers,[3a] distills at 67–82° (10 mm.), and 91 g. (72%) of product is collected as a colorless liquid (Note 2).

B. *2α,4α-Dimethyl-8-oxabicyclo[3.2.1]oct-6-en-3-one* (**2**). A 1-l., three-necked, round-bottomed flask is fitted with a 100-ml. dropping funnel having a nitrogen inlet tube, a magnetic stirrer, a thermometer, and an efficient double-surface condenser carrying a nitrogen outlet tube connected to a bubbler and is placed on a combined hotplate–magnetic stirring unit in a heat-resistant glass dish to act as a water bath. Dry acetonitrile (200 ml.) (Note 3) is introduced into the flask, and 90 g. (0.6 mole) of dried powdered sodium iodide (Note 4) is added with vigorous stirring under a slow stream of nitrogen. When the stirring bar rotates steadily, 20 g. (0.314 g.-atom) of powdered copper bronze (Note 5) is added, followed by 30 ml. (28 g., 0.4 mole) of freshly-distilled furan (Note 6). The dropping funnel is then charged with 24.4 g. (0.1 mole) of the dibromoketone 1 in 50 ml. of dry acetonitrile, and this solution is rapidly added to the stirred reaction mixture (Note 7). On addition of the dibromoketone 1 to the flask, the temperature rises to 45–50°, and a characteristic oatmeal-colored precipitate forms. After about 2 hours the temperature begins to drop, and the reaction is maintained at 50–60° with the water bath for a total reaction time of 4 hours (Note 8).

The flask is cooled to 0° with crushed ice (Note 9), and 150 ml. of dichloromethane is added with stirring. The reaction mixture is then poured into a 2-l. beaker containing 500 ml. of water and 500 ml. of crushed ice. Material remaining in the flask is rinsed into the beaker with 10 ml. of dichloromethane, and the mixture is stirred thoroughly, further salts being precipitated, until the ice just melts. The mixture is then filtered into a cooled filter flask under reduced pressure through a sintered or Büchner funnel and a kieselguhr filter-aid cake (Note 10). The beaker and filter cake are washed with 50 ml. of dichloromethane, and the clear combined filtrate is transferred to a 2-l. separatory funnel while still cold (Note 11).

The mixture is shaken vigorously, and the lower layer is separated and stored in ice. The aqueous layer is extracted with two 50-ml. portions of dichloromethane. The organic extracts are combined and shaken with 100 ml. of ice-cold, concentrated aqueous ammonia (35% w/w), and this mixture is filtered through a filter-aid cake and separated (Note 12). The extraction and filtration are repeated with fresh ammonia solution using the same filter (Note 13). The filter is washed with 50 ml. of dichloromethane, and the organic layer is separated and dried over anhydrous magnesium sulfate. The dried solution is filtered, the filter is washed with 50 ml. of dichloromethane, and the solvent is removed on a rotary evaporator at 30°. The flask containing the residual oil after removal of the dichloromethane is cooled to 0° before exposure to air (Note 14).

The light yellow oil is dissolved in 60 ml. of 30% anhydrous ethyl ether in pentane, and the solution is treated with 2 g. of anhydrous sodium sulfate and 0.5 g. of decolorizing carbon. The mixture is swirled for a few minutes, allowed to settle, and filtered through three sheets of fine filter paper by gravity into a 100-ml. round-bottomed flask with a 14/20 joint. The filter is washed with 10 ml. of pentane, and the flask is sealed by wiring on a 14-mm. serum cap. The flask is placed on a cork ring and lowered into an insulated container (large Dewar bottle, styrofoam box, etc.) half filled with dry ice, and is cooled slowly to −78°. When recrystallization is complete, a nitrogen supply is connected to the flask via a syringe needle; the supernatant liquid is then withdrawn by syringe and replaced with 50 ml. of pentane, previously cooled to −78°.

The flask is swirled to wash the crystals (Note 15), and the pentane is withdrawn. The flask is connected to a vacuum (water pump) via the nitrogen inlet and warmed to room temperature. The crude cycloadduct 2 (6.1–7.3 g., 40–48%) is isolated as colorless needles, m.p. 43.5–45°, from the first recrystallization (Note 16). Pure 2α,4α-dimethyl-8-oxabicyclo[3.2.1]oct-6-en-3-one (2) can be obtained by recrystallization from pentane at −78° with minimal loss, m.p. 45–46° (Note 17).

2. Notes

1. 3-Pentanone is available from Aldrich Chemical Company, Inc.

2. A slight coloration has no effect on the yield of the subsequent reactions. The dibromoketone 1 should be stored cold in a well-stoppered bottle (dark) under nitrogen, and is best handled cold to minimize spread of lachrymator vapors.

3. Commercial acetonitrile from Hopkins and Williams (GPR grade, given analysis 0.1% water, 0.02% acid) was used. Further purification of the solvent had no effect on the yield. The checker used MCB reagent-grade acetonitrile, refluxed over and distilled from calcium hydride.

4. The sodium iodide is dried at 150° for at least 3 hours, cooled in a desiccator, and finely powdered in a mortar. Although the use of less sodium iodide (e.g., 33 g., 0.22 mole) gives similar yields, the separation of the aqueous phase on extraction with dichloromethane and the following work-up are easier under the given conditions.

5. The copper bronze was supplied by BDH Chemicals Ltd. (Poole, England) as an extremely fine powder. The use of more granular electrolytic copper had no effect on the yield, but made magnetically operated stirring more difficult.

6. Commercial furan was refluxed over and distilled from calcium hydride and anhydrous potassium carbonate prior to use. Furan, b.p. 31°, is more volatile than ethyl ether, and precautions must be taken to minimize losses through evaporation.

7. Addition of dibromoketone 1 has been carried out all at once, slowly over a period of half an hour, and over 1 hour with no apparent change in yield.

8. Although a filtered sample of the reaction mixture analyzed by nuclear magnetic resonance (chloroform-d) shows no more dihaloketone after a reaction time of 2 hours (excess of furan was discernible), the reaction mixture must be heated for an additional 2 hours to destroy traces of dihaloketone which make difficult the subsequent work-up and analytical thin-layer chromatography of the product mixture. When the reaction was carried out with less sodium iodide (33 g., 0.22 mole), the presence of diiodoketone in the final product was noted by formation of an iodine color and

rapid decomposition of the cycloadduct to a black solid; also, the pentane washings developed an iodine color on exposure to light.

9. Since the dihaloketones may induce the decomposition of the product, it is essential to cool the solution in ice before allowing entry of air. Otherwise, the oily cycloadduct becomes brown, and polymeric material has to be removed before crystallization by passage down a 2×5 cm. column of silica gel (impregnated with 12% silver nitrate solution and redried).

10. Filtration through Hopkins and Williams kieselguhr filter-aid cake speedily removes even colloidal copper halides and breaks up any emulsion.

11. If the temperature is allowed to rise above $0°$, the cycloadduct decomposes and the yellow copper compound present liberates blue-green copper(II) salts which make the work-up difficult.

12. The checker, using 15 ml. of acetonitrile instead of 10 ml. of dichloromethane for rinsing the reaction vessel, found that no solid formed and filtration was not necessary. In this case three extractions with ammonia solution are required before addition of more ammonia fails to produce a blue color.

13. Using the same filter ensures that the ammonia solution is saturated with halide salts which aid final separation.

14. The oil is essentially pure cycloadduct 2, but owing to traces of impurity, crystallization may be difficult at $0°$. Analytical thin-layer chromatography on alumina, using low-polarity solvents such as pentane or carbon tetrachloride, was not successful in the presence of traces of dihaloketone, although it gave high resolution with related compounds. When dehalogenation is complete, however, the resolution is restored (see also Note 9).

15. If product 2 appears to be colored at this point, the pentane solution can simply be warmed to dissolve the crystals and cooled to recrystallize.

16. Approximately 0.4 g. of product 2 remaining in the supernatant solution can be recovered by repeated recrystallizations from the ethyl ether–pentane solution; however, the presence of 10 impurities (thin-layer chromatography) makes this rather impractical.

17. Infrared (carbon tetrachloride) cm.$^{-1}$: 1721 (very strong); proton magnetic resonance (chloroform-d) δ (multiplicity, number of protons, assignment, coupling constant J in Hz.): 6.35 (broad

singlet, 2, olefinic proton), 4.86 (doublet, bridgehead protons, 2, $J = 5$), 2.8 (doublet of quartets, 2, methine protons α to carbonyl, $J = 5$, $J = 7$), 0.98 (doublet, 6, methyl protons, $J = 7$).[3b]

3. Discussion

The reaction of 1,3-dibromo-1,3-diphenyl-2-propanone with sodium iodide in the presence of furan and cyclopentadiene to give bridged seven-membered rings has been reported by Cookson,[4a] who followed up earlier work by Fort.[4b] Chidgey[5a,b] demonstrated that the reaction can be extended to simple dibromoketones such as 1 and improved by using metallic copper to remove molecular iodine liberated during the reaction. Mechanistically, the reaction seems to involve two very fast S_N2 displacements of bromide by iodide ions, as seen by the precipitation of sodium bromide. A subsequent slower nucleophilic attack of excess iodide ion on the positively polarized iodine of the diiodoketone is visualized to yield an allylic iodide which forms an allyl cation in a fairly facile S_N1 reaction. The allyl cation is trapped by furan to give the oxygen-bridged seven-membered ring.[5b] The secondary–tertiary dibromoketone $CH_3CHBrCOCBr(CH_3)_2$ can also be used as an allyl cation precursor in this reaction, but the yields with the primary–tertiary dibromoketone $CH_2BrCOCBr(CH_3)_2$ are less satisfactory. The zinc-induced cycloaddition works well[3c] in this instance and also in the case of the ditertiary dibromoketone $(CH_3)_2CBrCOCBr(CH_3)_2$, which fails to undergo the initial S_N2 displacement with sodium iodide. Hence, although the sodium iodide–copper procedure is probably less general than the zinc[3b] and silver ion–promoted[5c] cycloaddition, it is experimentally convenient and yields preferentially the thermodynamically more stable adduct via an allyl cation in a W-configuration and compact transition state.[5b] The reaction involves inexpensive starting materials, proceeds under homogeneous conditions, and can be scaled-up readily.

The procedure described here has been used for the preparation of sensitive 6,7-dehydrotropinones in modest to good yields.[6] If in the present experiment furan is replaced by cyclopentadiene, an epimeric mixture of cis-diequatorial and cis-diaxial 2,4-dimethylbicyclo[3.2.1]oct-6-en-3-one is formed in almost 90%

$$CH_3CHBrCOCHBrCH_3 \xrightarrow[S_N2 \text{ (very fast)}]{2NaI} CH_3CHICOCHICH_3 + 2NaBr$$

$$\downarrow \begin{matrix} NaI \\ 2Cu \end{matrix}$$

yield.[7] Instead of zinc or sodium iodide/copper, nonacarbonyl-diiron may also be used as a reducing agent for dibromoketones.[8] Recently, $2\alpha,4\alpha$-dimethyl-8-oxabicyclo[3.2.1]oct-6-en-3-one (**2**) has been used as a precursor for the synthesis of ($\pm$)-nonactic acid, the building block of the macrotetrolide antibiotic nonactin.[9]

1. Chemistry Department, University College, London WC1H OAJ, England. Present address: Institut für Organische Chemie der Technischen Universität, Schneiderberg 1B, D-3000 Hannover 1, Germany. This work was supported by the Science Research Council and by the Petroleum Research Fund, administered by the American Chemical Society.
2. See A. I. Vogel, "A Text-Book of Practical Organic Chemistry," 3rd ed., Longmans, London, 1959, p. 91. The all-glass Dufton column is a plain tube into which a solid glass spiral, wound around a central tube or rod, is placed. The spiral should fit tightly inside the tube to prevent leakage of vapor between the walls of the column and the spiral.
3. (a) H. M. R. Hoffmann and J. G. Vinter, *J. Org. Chem.*, **39**, 3921 (1974); (b) H. M. R. Hoffmann, K. E. Clemens, and R. H. Smithers, *J. Amer. Chem. Soc.*, **94**, 3940 (1972).
4. (a) R. C. Cookson, M. J. Nye, and G. Subrahmanyam, *J. Chem. Soc. C*, 473 (1967); (b) A. W. Fort, *J. Amer. Chem. Soc.*, **84**, 2620, 4979 (1962).
5. (a) R. Chidgey, Ph.D. Thesis, University of London, 1975; (b) H. M. R. Hoffmann, *Angew. Chem. Int. Ed. Engl.*, **12**, 819 (1973); (c) H. M. R. Hoffmann, D. R. Joy, and A. K. Suter, *J. Chem. Soc. B*, 57 (1968); R. Schmid and H. Schmid, *Helv. Chim. Acta*, **57**, 1883 (1974); H. Mayr and B. Grubmüller, *Angew. Chem.*, **90**, 129 (1978).
6. G. Fierz, R. Chidgey, and H. M. R. Hoffmann, *Angew. Chem. Int. Ed. Engl.*, **13**, 410 (1974).
7. D. I. Rawson, unpublished work; A. Busch and H. M. R. Hoffmann, *Tetrahedron Lett.*, 2379 (1976).
8. R. Noyori, S. Makino, T. Okita, and Y. Hayakawa, *J. Org. Chem.*, **40**, 806 (1975); R. Noyori, Y. Baba, S. Makino, and H. Takaya, *Tetrahedron Lett.*, 1741 (1973).
9. M. J. Arco, M. H. Trammell, and J. D. White, *J. Org. Chem.*, **41**, 2075 (1976).

Appendix
Chemical Abstracts Nomenclature (Collective Index Number; Registry Numbers)

$2\alpha,4\alpha$-Dimethyl-8-oxabicyclo[3.2.1]oct-6-en-3-one: 8-Oxabicyclo-[3.2.1]oct-6-en-3-one, 2,4-dimethyl- (8,9); (*endo,endo*) (37081-58-6); (*exo,exo*) (37081-59-7)

3-Pentanone, 2,4-dibromo- (8,9); (815-60-1)

Furan (8, 9); (110-00-9)

2-Propanone, 1,3-dibromo-1,3-diphenyl- (8,9); (958-79-2)

Cyclopentadiene: 1,3-Cyclopentadiene (8,9); (542-92-7)

$CH_3CHBrCOCBr(CH_3)_2$: 3-Pentanone, 2,4-dibromo-2-methyl- (8,9); (37010-00-7)

$CH_2BrCOCBr(CH_3)_2$: 2-Butanone, 1,3-dibromo-3-methyl- (8,9); (1518-06-5)

$(CH_3)_2CBrCOCBr(CH_3)_2$: 3-Pentanone, 2,4-dibromo-2,4-dimethyl- (8,9); (17346-16-6)

(+)-Nonactic acid: 2-Furanacetic acid, tetrahydro-5-(2-hydroxypropyl)-α-methyl- (8); 2-Furanacetic acid, tetrahydro-5-(2-hydroxypropyl)-α-methyl-, [$2\alpha(R^*)$, $5\alpha(S^*)$]- (9); (60761-12-8)

BORANES IN FUNCTIONALIZATION OF DIENES TO CYCLIC KETONES: BICYCLO[3.3.1]NONAN-9-ONE

A. $(C_2H_5)_3COH \xrightarrow[\text{hexane, } 0°]{C_4H_9L:} (C_2H_5)_3COLi$

1

B.

9-BBN

1. CH_3OCHCl_2
2. $(C_2H_5)_3COLi$ (**1**)
tetrahydrofuran, $0°$

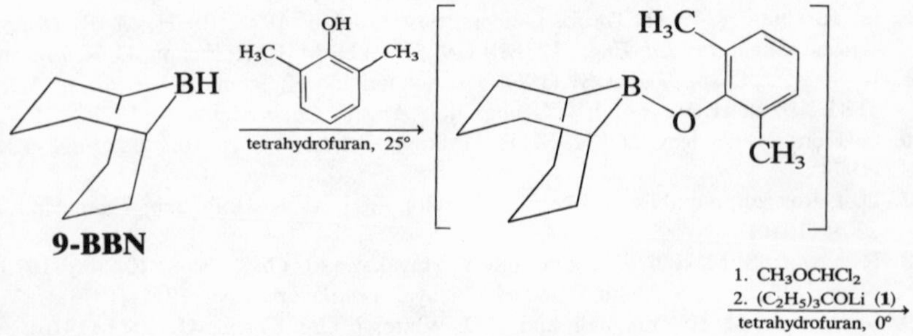

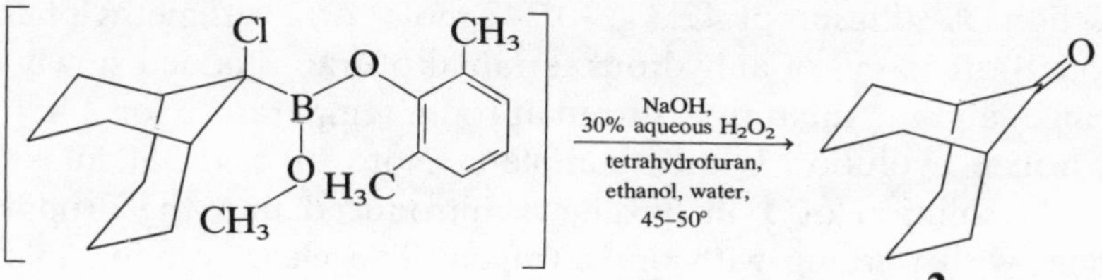

Submitted by BRUCE A. CARLSON[1] and HERBERT C. BROWN[2]
Checked by J. C. BOTTARO and G. BÜCHI

1. Procedure

Caution! The oxidation with 30% hydrogen peroxide in the last step of Part B may become vigorous and exothermic. No difficulties have been encountered under the conditions described; however, the oxidation should be carried out in a hood behind a protective shield.

A. *Lithium triethylcarboxide* (Solution **1**).   A thoroughly dried, 1-l., three-necked, round-bottomed flask is fitted with a septum inlet with serum cap, a reflux condenser, and a magnetic stirrer. The flask is purged with and maintained under an atmosphere of dry nitrogen. A solution of 300 ml. of 1.66 M butyllithium (0.50 mole) in hexane (Note 1) is introduced into the flask by syringe (Note 2) and cooled to 0° in an ice bath. Then 58 g. (0.50 mole) of 3-ethyl-3-pentanol (Note 3) is added slowly but constantly by syringe (Note 4). At the end of addition the yellow tint of the butyllithium solution disappears. The alkoxide solution is standardized by hydrolysis of aliquots in water and by titration of the resulting lithium hydroxide with standard acid to a phenolphthalein end point (Note 5).

B. *Bicyclo[3.3.1]nonan-9-one* (**2**).   A thoroughly dried 3-l., three-necked, round-bottomed flask is fitted with a serum cap and a magnetic stirrer under nitrogen. The flask is charged with 42.3 g. (0.347 mole) of 9-borabicyclo[3.3.1]nonane (9-BBN) (Note 6). Then 500 ml. of anhydrous tetrahydrofuran (Note 7) is added using a double-ended syringe needle (Note 2). The flask is fitted with a dry 500-ml. pressure-equalizing dropping funnel while the flask is purged with a rapid stream of dry nitrogen. The apparatus is maintained under a static pressure of nitrogen throughout the reaction. A solution of 42.3 g. (0.347 mole) of 2,6-dimethylphenol (Note 8) in 75 ml. of anhydrous tetrahydrofuran is added slowly by syringe, and the mixture is stirred at room temperature for 3 hours. Hydrogen evolution is then complete (Note 9), and 350 ml. of a 1.46 M solution of 1 in hexane is introduced into the dropping funnel with purging with dry nitrogen. The clear solution in the flask is cooled to 0°, and 44 g. (35 ml., 0.38 mole) of dichloro-methyl methyl ether (Note 10) is added. Solution 1 is then added slowly over approximately 30 minutes. The ice bath is removed, and the flask is warmed to room temperature for 90 minutes (Note 11). A heavy white precipitate of lithium chloride forms (Note 12), and the nitrogen atmosphere is no longer re-quired. A solution of 300 ml. of 95% ethanol, 70 ml. of water, and 42 g. of sodium hydroxide is added, and the mixture is cooled to 0° with efficient stirring. Oxidation is carried out by the slow, drop-wise addition of 70 ml. of 30% hydrogen peroxide (Note 13), while the temperature is maintained below 50° with a cooling bath. The addition is carried out over 40–45 minutes. After addition the mixture is heated with stirring to 45–50° for 2 hours (Note 14) and then cooled to room temperature. Water (300 ml.) is added, and the aqueous phase is saturated with sodium chloride. The organic phase is separated and washed with 100 ml. of saturated aqueous sodium chloride. The solvents are removed on a rotary evaporator, and the resulting orange liquid is diluted with 500 ml. of pentane. The pentane solution is extracted with 250-ml. and 100-ml. por-tions of 3 M aqueous sodium hydroxide to remove the 2,6-dimethylphenol. After washing with 100 ml. of saturated aqueous sodium chloride, the pentane is removed on a rotary evaporator. The 3-ethyl-3-pentanol is next removed by distillation under a water aspirator vacuum, b.p. 54–56° (16 mm.) (Note 15). The resulting semisolid residue is dissolved in 200 ml. of pentane, filtered to remove the impurities, and the ketone 2 is crystallized by cooling the filtrate to −78°. The ketone 2 is collected by suction filtration, washed with 50 ml. of −78° pentane (Note 16), and dried (Note 17). A yield of 35–40 g. (78–83%) of pure ketone 2 is obtained, m.p. 154–156° (Note 18). Evaporation of the filtrate to *ca.* 50 ml. and cooling to −78° gives another 3–4 g. of ketone 2, m.p. 149–154°. The total yield of bicyclo[3.3.1]nonan-9-one (2) is 38–44 g. (79–91%).

2. Notes

1. Butyllithium in hexane was obtained from Foote Mineral Company. The solution can be titrated for total alkyllithium by the procedure of Watson and Eastham,[5] and for total base by hydrolysis of aliquots and titration against standard acid prior to use. Butyllithium solution of good purity is essential for success of the reaction.

2. A double-ended syringe needle is most convenient for transfer of reagents and handling of air-sensitive solutions. Transfer is completed by applying nitrogen pressure. See references 3 and 4 for descriptions of this procedure and for general techniques for handling air-sensitive reagents.

3. 3-Ethyl-3-pentanol (triethylcarbinol) (98%) was purchased from Chemical Samples Company, 469 Kenny Road, Columbus, Ohio 43221, and used without further purification. It is also available from Aldrich Chemical Company, Inc.

4. Addition causes vigorous reaction with evolution of butane and heat. Redissolution of the butane in the solution may create a partial vacuum, causing air to be sucked back into the flask. The addition should be kept at such a rate to prevent this. The solution may reflux, but this causes no harm.

5. Alternatively, if only one use of the solution of lithium triethylcarboxide (**1**) is planned, the required amounts of butyllithium solution and 3-ethyl-3-pentanol may be reacted, and the resulting solution of lithium triethylcarboxide used without standardization.

6. 9-BBN was purchased from Aldrich Chemical Company, Inc., and used directly. Alternatively, 9-borabicyclo[3.3.1]nonane may be prepared by hydroboration of 1,5-cyclooctadiene.[4] *Caution! 9-BBN is air sensitive and can be pyrophoric.*

7. Tetrahydrofuran, containing less than 0.002% peroxides, supplied by the William A. Mosher, Ph.D. Company, 101 Townsend Road, Newark, Delaware 19711, was used directly. Tetrahydrofuran from other sources may require purification prior to use. Stirring with lithium aluminum hydride or sodium benzophenone ketyl, followed by distillation under an inert atmosphere, is the recommended procedure for removal of water and peroxides (see reference 4, p. 256).

8. 2,6-Dimethylphenol (99%) from Chemical Samples Company was used without purification. It is also available from Aldrich Chemical Company, Inc.

9. Addition causes evolution of hydrogen and is carried out over *ca.* 10 minutes to prevent foaming. The undissolved 9-BBN rapidly goes into solution. Completion of the reaction may be measured by monitoring hydrogen evolution with a gas meter. The theoretical amount of hydrogen is 7.8 l. at STP.

10. Dichloromethyl methyl ether (98%) was purchased from Aldrich Chemical Company, Inc. (listed as α,α-dichloromethyl methyl ether), and used without purification. Better results can be obtained if the material is distilled under nitrogen prior to use. Unlike chloromethyl methyl ether and bis(chloromethyl) ether, dichloromethyl methyl ether is reported to have no significant carcinogenic activity.[6] However, as a precaution it should be handled carefully in a well-ventilated hood.

11. Longer reaction times (*e.g.*, allowing the reaction to stir overnight) will not affect the product.

12. The checkers observed no precipitate in their two runs. On occasion the submitters have also noted no lithium chloride precipitate during a run or sudden precipitation of the salt. Apparently the lithium chloride supersaturates on these occasions.

13. The 30% hydrogen peroxide was obtained from Fisher Scientific Company.

14. Initial heating should be done cautiously in case an additional exotherm is experienced.

15. The 3-ethyl-3-pentanol may be removed rapidly by condensation into a dry ice–acetone cold trap.

16. The temperature of the solution should be kept as close as possible to $-78°$ during filtration to prevent losses of the first crop of product. On humid days crystallization and filtration are best accomplished in a closed system to prevent excessive condensation of water.

17. The product has a high vapor pressure and sublimes readily. Care should be taken in drying so that losses of product are not incurred. Drying briefly in a vacuum desiccator over Drierite is recommended.

18. Literature[7] m.p. 155-158.5°; the 2,4-dinitrophenyl hydrazone was prepared by standard procedure and recrystallized

from ethyl acetate–ethanol to provide orange prisms, m.p. 190–191.2°, literature[7] m.p. 191.8–192.3°.

3. Discussion

Dialkylborinic acids are conveniently prepared by the hydroboration of olefins with monochloroborane diethyletherate[4c] or, in some cases, by controlled hydroboration of two equivalents of olefin[4c] with borane–tetrahydrofuran or borane–methyl sulfide complex followed by alcoholysis. The base-induced reaction of dichloromethyl methyl ether with borinic acid esters provides a route to a variety of ketones[8] and olefins.[9,10] The reaction presumably proceeds via generation of a haloalkoxy carbanion or carbene which reacts rapidly with the borinic acid ester, resulting in transfer of the alkyl groups from boron to carbon. The α-chloroboronic ester intermediate thus generated is an exceptionally versatile intermediate. Oxidation results in the formation of the corresponding ketone in high yield,[8,12] whereas pyrolysis or solvolysis with aqueous silver nitrate[10] provides the internal olefin:

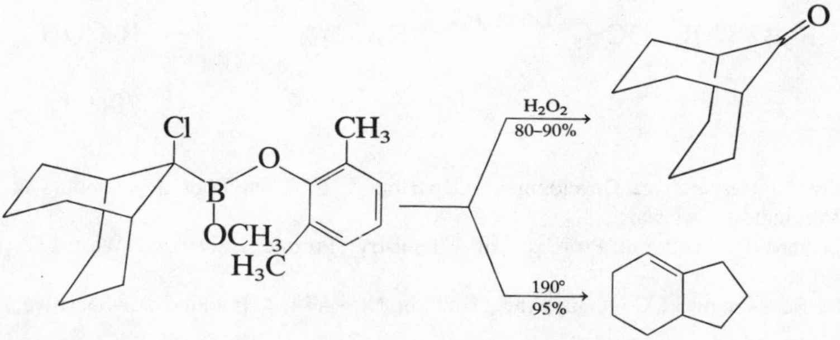

Lithium triethylcarboxide (**1**) is the base of choice for these conversions. The use of less hindered alkoxide or amide bases results in poorer yields. In this procedure 1.5–2.0 equivalents of base are required, although with more bulky alkyl groups attached to boron only 1 equivalent is necessary. The use of the more hindered 2,6-diisopropylphenol to form the borinic ester gives a 96% yield of the bicyclic ketone with only 1 equivalent of base; however, in the work-up procedure this phenol is more difficult to separate from the ketone.

Carbonylation of trialkylboranes in the presence of water[4e] and the cyanation of thexyldialkylboranes[11] offer alternative routes to ketones from organoboranes. Neither procedure can utilize dialkylborinic esters in the synthesis and, thus, result in loss of one alkyl group. Carbonylation requires a moderate pressure of carbon monoxide in some cases, and the cyanation reaction involves the use of strongly electrophilic reagents; neither route has been used successfully in the preparation of bicyclo[3.3.1]nonan-9-one (2). Previous synthetic routes to this interesting bicyclic ketone[13] have generally required numerous steps and resulted in low overall yields.[7,14] The best alternative procedure, which gives a 60% yield of the bicyclic ketone, involves reaction of nickel carbonyl with 1,5-cyclooctadiene.[15]

The base-induced reaction of dichloromethyl methyl ether with trialkyl- and triarylboranes also provides a powerful method for the preparation of the corresponding tertiary carbinols.[16,17] In this case, all three groups transfer readily from boron to carbon under mild conditions, and oxidation with alkaline hydrogen peroxide provides the tertiary alcohol:

$$R_3B + CHCl_2OCH_3 \xrightarrow{\text{LiOC(C}_2\text{H}_5)_3} R_3C-B\begin{matrix} \diagup Cl \\ \diagdown OCH_3 \end{matrix} \xrightarrow{[O]} R_3COH$$

70–90%

1. Central Research & Development Department, E. I. du Pont de Nemours & Co., Wilmington, Delaware.
2. Richard B. Wetherill Professor of Chemistry, Purdue University, West Lafayette, Indiana.
3. Aldrich Chemical Company, Inc., Bulletin No. A74, "Handling Air-Sensitive Solutions."
4. H. C. Brown, "Organic Syntheses via Boranes," John Wiley & Sons, New York, N. Y., 1975; (a) pp. 191–261; (b) pp. 32–34; (c) pp. 45–47; (d) pp. 3–9; (e) pp. 127–130.
5. S. C. Watson and J. F. Eastham, J. Organometal. Chem., 9, 165 (1967). The procedure is described in detail in reference 4, pp. 248–249.
6. B. L. Van Duuren, C. Katz, B. M. Goldschmidt, K. Frenkel, and A. Sivak, J. Nat. Cancer Inst., U.S.A., 48, 1431 (1972).
7. C. S. Foote and R. B. Woodward, Tetrahedron, 20, 687 (1964).
8. B. A. Carlson and H. C. Brown, J. Amer. Chem. Soc., 95, 6876 (1973).
9. J. J. Katz, B. A. Carlson, and H. C. Brown, J. Org. Chem., 39, 2817 (1974).
10. H. C. Brown, J. J. Katz, and B. A. Carlson, J. Org. Chem., 40, 813 (1975).
11. A. Pelter, K. Smith, M. G. Hutchings, and K. Rowe, J. Chem. Soc., Perkin Trans. 1, 129 (1975).
12. B. A. Carlson and H. C. Brown, Synthesis, 776 (1973).

13. A review of structure and reactivities in the bicyclo[3.3.1]nonane system is provided by H. Kato in *J. Synth. Org. Chem. Jap.*, **28**, 682 (1966).
14. R. D. Allan, B. G. Cordiner, and R. J. Wells, *Tetrahedron Lett.*, 6055 (1968).
15. B. Fell, W. Seide, and F. Asinger, *Tetrahedron Lett.*, 1003 (1968).
16. H. C. Brown and B. A. Carlson, *J. Org. Chem.*, **38**, 2422 (1973).
17. H. C. Brown, J. J. Katz, and B. A. Carlson, *J. Org. Chem.*, **38**, 3968 (1973).

Appendix
Chemical Abstracts Nomenclature (Collective Index Number; Registry Numbers)

Bicyclo[3.3.1]nonan-9-one (8,9); (17931-55-4)

Lithium triethylcarboxide: 3-Pentanol, 3-ethyl-, lithium salt (8,9); (32777-93-8).

Lithium, butyl- (8,9); (109-72-8)

3-Pentanol, 3-ethyl- (8,9); (597-49-9)

9-Borabicyclo[3.3.1]nonane (8,9); (280-64-8)

2,6-Xylenol (8); Phenol, 2,6-dimethyl- (9); (576-26-1)

Ether, dichloromethyl methyl (8); Methane, dichloromethoxy- (9); (4885-02-3)

1,5-Cyclooctadiene (8,9); (111-78-4)

Lithium aluminum hydride: Aluminate (1-), tetrahydro-, lithium (8); Aluminate (1-), tetrahydro-, lithium, (T-4)- (9); (16853-85-3)

Sodium benzophenone ketyl: Benzophenone, radical ion (1-), sodium (8); Methanone, diphenyl-, radical ion (1-), sodium (9); (3463-17-0)

Ether, chloromethyl methyl (8); Methane, chloromethoxy- (9); (107-30-2)

Ether, bis(chloromethyl)- (8); Methane, oxybis[chloro- (9); (542-88-1)

Bicyclo[3.3.1]nonan-9-one, (2,4-dinitrophenyl)hydrazone (8,9);(−)

Monochloroborane diethyletherate: Borane, chloro-, compound with ethyl ether (8); Borane, chloro-, compound with 1,1'-oxybis[ethane] (1:1) (9); (36594-41-9).

Borane–methyl sulfide: Borane, compound with methyl sulfide (1:1) (8); Borane, compound with thiobismethane (1:1) (9); (13292-87-0)

Nickel carbonyl [Ni(CO)$_4$] (8,9); (13463-39-3)

BORANES IN FUNCTIONALIZATION OF OLEFINS TO AMINES: 3-PINANAMINE

(Bicyclo[3.1.1]heptan-3-amine, 2,6,6-trimethyl-)

A. $(NH_2OH)_2 \cdot H_2SO_4 + 2ClSO_3H \longrightarrow \overset{+}{N}H_3OSO_3^- + 2HCl + H_2SO_4$

1

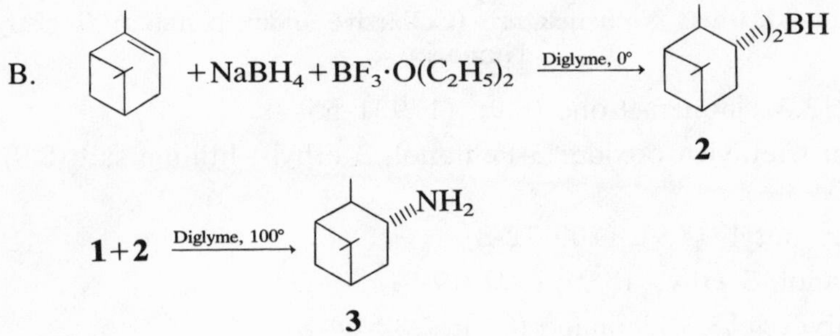

B.

2

3

Submitted by MICHAEL W. RATHKE[1] and ALAN A. MILLARD
Checked by ARNOLD BROSSI and JUN-ICHI MINAMIKAWA

1. Procedure

A. *Hydroxylamine-O-sulfonic acid* (**1**). A 500-ml., three-necked, round-bottomed flask is fitted with a mechanical stirrer, dropping funnel, and drying tube with calcium chloride. Finely powdered hydroxylamine sulfate (26.0 g., 0.16 mole) (Note 1) is placed in the flask, and 60 ml. (107.4 g., 0.92 mole) of chlorosulfonic acid (Note 1) is added dropwise over 20 minutes with vigorous stirring (Note 2). After the addition is complete, the flask, with stirring, is placed in a 100° oil bath for 5 minutes. The pasty mixture is cooled to room temperature, and the flask is placed in an ice bath. To the stirred mixture 200 ml. of ethyl ether is slowly added over 20–30 minutes (Note 3). During the ethyl ether addition, the pasty contents change to a colorless powder which is collected by suction on a Büchner funnel. The powder is washed with 300 ml. of tetrahydrofuran, then with 200 ml. of ethyl ether. The product **1**, after drying, weighs 34–35 g. (95–97%). Iodometric titration shows the product is 96–99% pure (Note 4), which is adequate for use in the following reaction.

32

B. 3-*Pinanamine* (**3**). A 1000-ml., three-necked, round-bottomed flask is fitted with a gas-inlet tube, a reflux condenser connected to a mineral oil bubbler, and a sealed mechanical stirrer. The system is flamed with a Bunsen burner while being flushed with dry nitrogen. The reaction vessel is then cooled under a nitrogen stream in an ice bath while a slight positive pressure of nitrogen is maintained. A solution of 3.12 g. (0.0824 mole) of sodium borohydride (Note 5) in 100 ml. of diglyme (Note 6) is added to the flask, followed by 27.25 g. (0.20 mole) of ($\pm$)-α-pinene (Note 7). Hydroboration is achieved by dropwise addition of 15.6 g. (0.11 mole) of boron trifluoride etherate (Note 8) over a 15-minute period. Di-3-pinanylborane (**2**) precipitates as a white solid. The ice bath is removed, and the reaction mixture is stirred at room temperature for 1 hour. Hydroxylamine-*O*-sulfonic acid (**1**) (24.9 g., 0.22 mole) in 100 ml. of diglyme is added dropwise to the mixture over a 5-minute period (Note 9). The mixture is then heated in a 100° oil bath for 3 hours. The mixture is cooled to room temperature, and 80 ml. of concentrated hydrochloric acid is added over a 5-minute period. The mixture is poured into 800 ml. of water and extracted with two 100-ml. portions of ethyl ether. The ether layers are discarded, and the aqueous layer is made alkaline with sodium hydroxide pellets (60–65 g. is needed). The aqueous layer is extracted with two 100-ml. portions of ethyl ether, the combined ether extracts are dried over anhydrous sodium sulfate, and the drying agent is removed by filtration. The filtrate is transferred to a 500-ml., ice-cooled flask fitted with a magnetic stirring bar. A solution of 85–88% phosphoric acid (12 g., 0.10 mole) in 100 ml. of ethanol is added to the flask over 10 minutes with stirring. The precipitated colorless crystals are collected by suction on a Büchner funnel, and the salt is suspended in 300 ml. of hot water contained in a 1-l. flask. The mixture is heated with magnetic stirring in a 120–130° oil bath until all the salt has dissolved (*ca.* 20–30 minutes), then quickly filtered by suction. Pure phosphate salt immediately precipitates as colorless plates, which are collected on a Büchner funnel and dried in a desiccator. The yield is 16.6 g. (33.1%). A second crop of 4.4 g. can be obtained by concentrating the mother liquor to about half its original volume. The total yield of pure phosphate salt is 21.0 g. (41.8%), m.p. 275–280° (dec.) (Note 10). The salt is easily converted to free amine **3** by the following procedure: 10 g. (0.04 mole) of the salt is dissolved in 40 ml. of aqueous 3 M sodium hydroxide and extracted with two 50-ml. portions of ethyl ether. The combined extracts are dried over anhydrous sodium sulfate, the drying agent is removed by filtration, and the solvent is removed under reduced pressure by a rotary evaporator. The residual oil is distilled to give 5.9 g. (93% from phosphate salt) of amine **3** as a colorless liquid, b.p. 83° (13 mm.) (Note 11).

2. Notes

1. Commercial hydroxylamine sulfate and chlorosulfonic acid, obtained from Eastman Kodak Company, were used directly. The checkers found that commercially available hydroxylamine-O-sulfonic acid is sometimes of low purity. Therefore, the use of freshly prepared reagent is recommended.

2. Hydrogen chloride gas is evolved during the addition. The reaction should be carried out in a hood. An aqueous base scrubber is also recommended.

3. Rapid addition of the ethyl ether must be avoided because of its high reactivity with chlorosulfonic acid.

4. Iodometric titration was carried out as shown below: About 100 mg. of hydroxylamine-O-sulfonic acid was exactly weighed and dissolved in 20 ml. of distilled water. Sulfuric acid (10 ml. of 10% solution) and 1 ml. of saturated potassium iodide solution were then added. After the solution was allowed to stand for 1 hour, liberated iodine was titrated with 0.1N sodium thiosulfate solution until the iodine color disappeared. The following stoichiometric relation was used: 0.1N $Na_2S_2O_3$ (1 ml.) = 5.66 mg. $H_3\overset{+}{N}OSO_3^-$. Hydroxylamine-O-sulfonic acid should be stored in tightly sealed bottles in a refrigerator.

5. Commercial sodium borohydride was obtained from Ventron Corporation and used directly.

6. Commercial diglyme (dimethyl ether of diethylene glycol) was obtained from Ansul Chemical Company, Marinette, Wisconsin, and purified by distillation from lithium aluminum hydride at 62–63° (15 mm.)[2].

7. ($\pm$)-α-Pinene, b.p. 54° (22 mm.), was obtained from Aldrich Chemical Company, Inc., and distilled before use.

8. Commercial boron trifluoride etherate, b.p. 46° (10 mm.), available from Matheson Coleman and Bell, was distilled from calcium hydride before use.

9. *Caution: Since* (±)-α-*pinene is hydroborated to the dialkylborane state* (R_2BH), *a large amount of hydrogen is evolved on addition of hydroxylamine-O-sulfonic acid. Consequently, the addition should be carried out dropwise and adequate ventilation should be provided.*

10. The phosphate salt corresponds to the empirical formula $C_{10}H_{19}N \cdot H_3PO_4$.

11. The product showed one peak on gas-phase chromatography (3% SE-30, 70°C).

3. Discussion

Hydroxylamine-*O*-sulfonic acid can also be prepared from hydroxylamine sulfate and 30% fuming sulfuric acid (oleum).[3] The present procedure is essentially that of F. Sommer *et al.*[4]

The hydroboration–amination sequence in diglyme is a general procedure for the conversion of olefins to primary amines without rearrangement and with predictable stereochemistry.[5] An alternative procedure, using tetrahydrofuran as solvent and either hydroxylamine-*O*-sulfonic acid or chloramine, is applicable with terminal olefins and relatively unhindered internal and alicyclic olefins.[6] *O*-Mesitylenesulfonylhydroxylamine also gave desired amines in comparable yield.[7] Alternative procedures for the hydroboration of olefins use commercially available solutions of diborane in tetrahydrofuran[8] or dimethylsulfide.[9]

Olefins may be converted to primary amines by the Ritter reaction[10] or by reaction with mercuric nitrate in acetonitrile solution.[11] In both cases regiospecificity for the formal addition of ammonia across the double bond is opposite to that observed in the hydroboration-amination sequence.

1. Department of Chemistry, Michigan State University, East Lansing, Michigan 44824.
2. G. Zweifel and H. C. Brown, *Org. Syn.*, **52,** 59 (1972).
3. H. J. Matsuguma and L. F. Audrieth, *Inorg. Syn.*, **5,** 122 (1957).
4. F. Sommer, O. F. Schulz, and M. Nassau, *Z. Anorg. Allg. Chem.*, **147,** 142 (1925).
5. M. W. Rathke, N. Inoue, K. R. Varma, and H. C. Brown, *J. Amer. Chem. Soc.*, **88,** 2870 (1966).

36 ORGANIC SYNTHESES—VOL. 58

6. H. C. Brown, W. R. Heydkemp, E. Breuer, and W. S. Murphy, *J. Amer. Chem. Soc.*, **86**, 3565 (1964).
7. Y. Tamura, J. Minamikawa, S. Fujii, and M. Ikeda, *Synthesis*, 196 (1974).
8. H. C. Brown, "Organic Synthesis via Boranes", John Wiley & Sons, New York, N. Y. (1975).
9. C. F. Lane, *J. Org. Chem.*, **39**, 1437 (1974).
10. L. I. Krimen and D. J. Cota, *Org. React.*, **17**, 213 (1969).
11. H. C. Brown and J. T. Kurek, *J. Amer. Chem. Soc.*, **91**, 5647 (1969).

Appendix

Chemical Abstracts Nomenclature (Collective Index Number; Registry Numbers)

3-Pinanamine (8); (17371-27-6); Bicyclo[3.1.1]heptan-3-amine, 2,6,6-trimethyl- (9); (1α,2α,3α,5α)- (35117-66-9); (1α,2α,3β,5α)- (35117-55-6); (1α,2β,3α,5α)- (35117-58-9); (1α,2β,3β,5α)- (35117-61-4)

Hydroxylamine-*O*-sulfonic acid (8,9); (2950-43-8)

Hydroxylamine, sulfate (2:1) (8,9); (10039-54-0)

Chlorosulfonic acid: Chlorosulfuric acid (8,9); (7790-94-5)

Sodium borohydride: Borate(1-), tetrahydro-, sodium (8,9); (16940-66-2)

Diglyme: Ether, bis(2-methoxyethyl) (8); Ethane, 1,1'-oxybis[2-methoxy- (9); (111-96-6)

(±)-α-Pinene: 2-Pinene (8); Bicyclo[3.1.1]-hept-2-ene, 2,6,6-trimethyl- (9); (80-56-8); (±) (2437-95-8)

Borane, di-3-pinanyl- (8); Borane, bis(2,6,6-trimethylbicyclo-[3.1.1]hept-3-yl)- (9); (1091-56-1)

Lithium aluminum hydride: Aluminate(1-), tetrahydro-, lithium (8); Aluminate(1-), tetrahydro-, lithium, (T-4)- (9); (16853-85-3)

Chloramine: Chloramide (8,9); (10599-90-3)

O-Mesitylene sulfonylhydroxylamine: Hydroxylamine, *O*-mesitylsulfonyl- (8); Hydroxylamine, *O*-[(2,4,6-trimethylphenyl)sulfonyl]- (9); (36016-40-7)

Diborane (4) (8,9); (18099-45-1)

Dimethyl sulfide: Methyl sulfide (8); Methane, thiobis- (9); (75-18-3)

CARBENE GENERATION BY α-ELIMINATION WITH LITHIUM 2,2,6,6-TETRAMETHYLPIPERIDIDE: 1-ETHOXY-2-p-TOLYL-CYCLOPROPANE

[Benzene, 1-(2-ethoxycyclopropyl)-4-methyl]

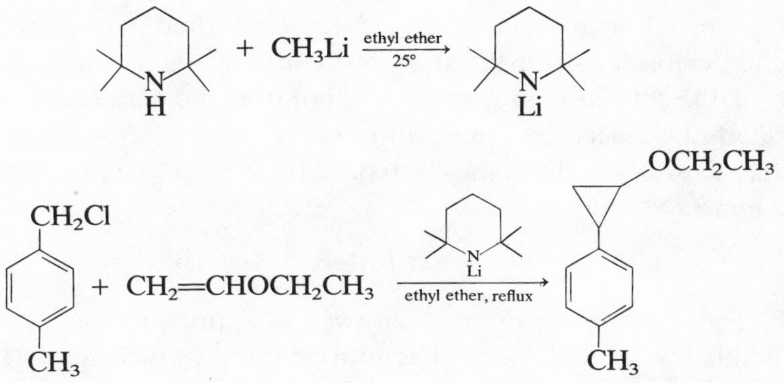

Submitted by Charles M. Dougherty[1] and Roy A. Olofson[2]
Checked by Mark W. Johnson and Robert M. Coates

1. Procedure

Caution! See benzene warning, p. 168.

A 250-ml., three-necked, round-bottomed flask is equipped with a 50-ml. pressure-equalizing dropping funnel capped by a rubber septum, an efficient reflux condenser connected to a nitrogen inlet, and a magnetic stirrer (Note 1). The flask is charged with 7.02 g. (0.05 mole) of α-chloro-p-xylene (Note 2) and 45.6 g. (0.63 mole) of ethyl vinyl ether (Note 3). A solution of 7.06 g. (0.05 mole) of 2,2,6,6-tetramethylpiperidine (Note 4) in 15 ml. of dry ethyl ether is injected through the septum into the dropping funnel. Lithium 2,2,6,6-tetramethylpiperidide is generated *in situ* by injecting 46.5 ml. (0.05 mole) of a 1.08 M solution of methyllithium in ethyl ether (Note 5) through the septum over a 5–10-minute period (Notes 6 and 7). After another 10 minutes, the contents are added dropwise to the vigorously stirred solution in the flask at a rate that maintains a gentle reflux. When the *ca.* 2-hour addition period is complete, the white slurry is allowed to stir overnight at room temperature (Note 8). Water (10 ml.) is added dropwise to the stirred suspension, and the contents of the flask are poured into a separatory funnel containing 100 ml. of ethyl ether and 100 ml. of water. The aqueous layer is separated and extracted with two more 100-ml. portions of ethyl ether. The combined ether solutions are washed with 10% aqueous citric acid (Note 9), 5% aqueous sodium bicarbonate, and water before being dried with anhydrous calcium chloride. After filtration of the drying agent and evaporation of the filtrate with a rotary evaporator, the residual liquid is distilled at reduced pressure, affording 6.6–7.0 g. (75–80%) of 1-ethoxy-2-p-tolylcyclopropane, b.p. 116–118° (10 mm.), 95–96° (3.2 mm.) (Note 10).

2. Notes

1. The glassware is dried in an oven at approximately 125° and assembled while still warm. The nitrogen inlet, which consists of a T-tube assembly connected to an oil bubbler, is attached, and the apparatus is allowed to cool while being swept with a stream of dry nitrogen. The septum is placed on top of the dropping funnel, and the nitrogen flow adjusted to maintain a slight positive pressure of nitrogen within the apparatus during the reaction.

2. α-Chloro-p-xylene was obtained from Aldrich Chemical Company, Inc., and purified by distillation under reduced pressure.

3. Ethyl vinyl ether was supplied by Aldrich Chemical Company, Inc., and distilled from sodium. If simple alkenes are used in place of ethyl vinyl ether, the submitters find that the yields of cyclopropanes are improved by dilution of the olefin with one or two volumes of ethyl ether.

4. 2,2,6,6-Tetramethylpiperidine, furnished by Aldrich Chemical Company, Inc., Fluka A G, and ICN Life Sciences Group, is sometimes contaminated with traces of water, hydrazine, and/or 2,2,6,6-tetramethyl-4-piperidone. These impurities may be. removed by drying with sodium hydroxide or potassium hydroxide pellets, filtering, and distilling at atmospheric pressure, b.p. 153–154°. The purified amine can be stored indefinitely under a nitrogen atmosphere.

5. Methyllithium in ethyl ether from Ventron Corporation was used. Directions for the preparation of ethereal methyllithium from methyl bromide are also available [see G. Wittig and A.

Hesse, *Org. Syn.*, **50,** 66 (1970)]. The checkers standardized the solution immediately before use by diluting a 2.5-ml. aliquot with 10 ml. of benzene and titrating with a 1 M solution of 2-butanol in xylene according to the procedure of Watson and Eastham[3] [see M. Gall and H. O. House, *Org. Syn.*, **52,** 39 (1972)], with 1,10-phenanthroline as indicator. The submitters report that the yield of arylcyclopropane is lower if a commercially available solution of butyllithium in hydrocarbon solvents is used.

6. The methane generated is vented by passage through the oil bubbler.

7. Since the reaction between methyllithium and 2,2,6,6-tetramethylpiperidine is relatively slow at lower temperatures, lithium 2,2,6,6-tetramethylpiperidide is best prepared at room temperature. The reagent may, however, be used over a wide range of temperatures.

8. Approximately the same yields are obtained if the product is isolated after 2–3 hours.

9. The use of aqueous citric acid avoids the formation of insoluble gelatinous precipitates, which result when aqueous hydrochloric acid is employed. Sulfuric acid is a suitable alternative to citric acid but must be used in substantial excess to prevent precipitation. 2,2,6,6-Tetramethylpiperidine may be recovered from the citric acid extract by making the aqueous solution basic and extracting with ethyl ether.

10. The product, a mixture of *cis-* and *trans*-isomers in the ratio of about 2 : 1, has the following spectral properties: infrared (liquid film) cm^{-1} (strong): 1510, 1440, 1370, 1340, 1120, 1080; proton magnetic resonance (carbon tetrachloride) δ (multiplicity, number of protons, assignment): 0.63–1.3 (multiplet, 5, cyclopropyl CH_2 and ethoxy CH_3), 1.4–2.0 (multiplet, 1, cyclopropyl CH), 2.23 (singlet, 1, *trans*-aromatic CH_3), 2.28 (singlet, *ca.* 2, *cis*-aromatic CH_3), 2.8–3.7 (multiplet, 3, cyclopropoxy CH and ethoxy CH_2), 6.7–7.2 (multiplet, 4, aromatic H.) The following specific absorptions in the proton magnetic resonance spectrum may be used to estimate the ratio of the two isomers, δ (multiplicity, number of protons, assignment, coupling constant J in Hz.): *cis*-isomer: 0.92 (triplet, 3, ethoxy CH_3, $J = 7$), 7.02 (center of $AA'BB'$ multiplet, 4, aromatic H); *trans*-isomer: 1.14 (triplet, 3, ethoxy CH_3, $J = 7$), 6.88 (center of $AA'BB'$ multiplet, 4, aromatic H).

3. Discussion

This procedure describes the generation of the strong non-nucleophilic amide base, lithium 2,2,6,6-tetramethylpiperidide. This base is used in the regioselective abstraction of a proton from a very weak carbon acid containing other sites reactive towards nucleophilic attack.[4,5] In contrast, most other strong bases undergo preferential alkylation with benzyl halides. 1-Ethoxy-2-*p*-tolylcyclopropane is one of over a dozen aryl cyclopropanes, cyclopropenes, and cyclopropanone ketals that have been prepared by this method[5] (Table I). An analog, 1-methoxy-2-phenyl-

TABLE I
PREPARATION OF CYCLOPROPANES FROM ALKYL HALIDES, ALKENES, AND
LITHIUM 2,2,6,6-TETRAMETHYLPIPERIDIDE

Product	Yield (%)	Product	Yield (%)
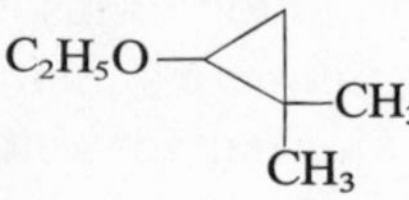	66[7]	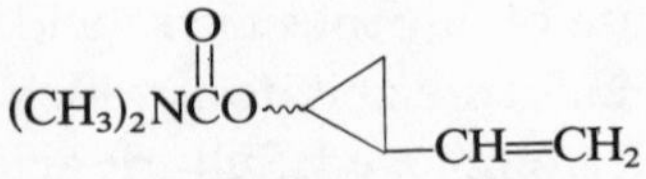	21[9]
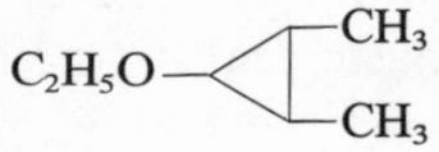	66[7]	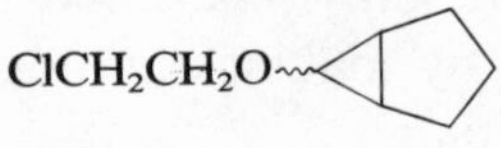	64[8]
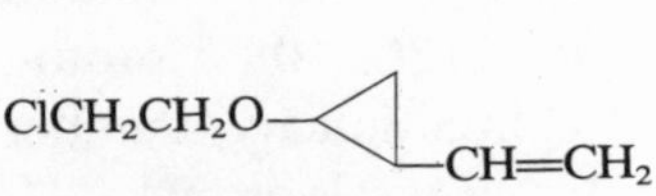	74[8]	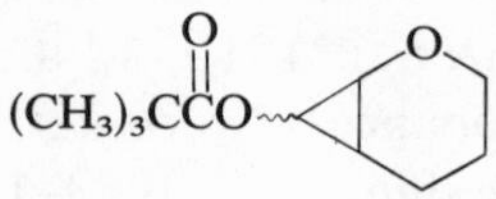	39[9]
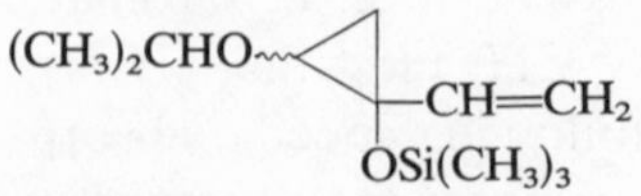	46[7]	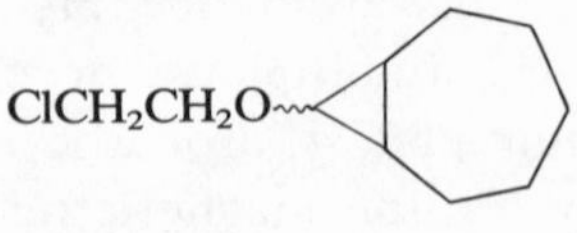	62[8]
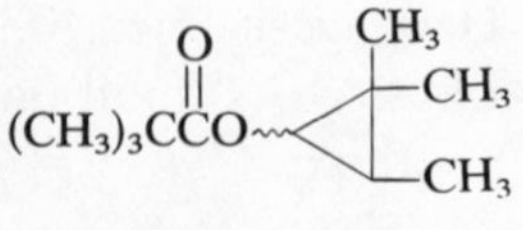	35[9]	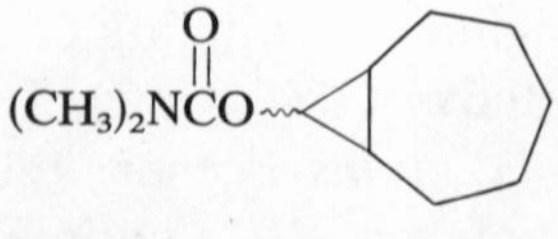	20 (24)[9]

cyclopropane, has been obtained in 8% yield from the reaction of methyllithium with dichloromethyl methyl ether in styrene.[6] The alkene is present in large excess, as is commonly the case for reactions involving short-lived carbene intermediates. In the present procedure ethyl vinyl ether serves as both solvent and reactant. For best results with alkenes lacking alkoxy substituents, approximately two volumes of ethyl ether or tetrahydrofuran should be used as diluent. Alkoxy,[7,8] acyloxy,[9] alkenyl,[5,10] trialkylsilyl,[10] and trialkylstannyl[10] carbenes have been generated and trapped *in situ* with alkenes and alkynes by this method, thus affording cyclopropanes with a variety of substituents.

Lithium 2,2,6,6-tetramethylpiperidide has also been used to advantage in a number of other types of reactions. This base reacts with aryl halides[5,10-12] (and, less cleanly, with aryl sulfonates[13]) to give benzynes, which have been trapped with thiolates,[5] acetylides,[5,13] enolates,[10,12,13] and conjugated dienes.[10,11,13] Replacement of halogen by hydrogen, a major reaction observed between other dialkylamide bases and aryl halides, does not occur with lithium 2,2,6,6-tetramethylpiperidide.[5] While alkyl benzoates undergo selective deprotonation in the ortho-position upon treatment with this amide base,[14] methyl thiobenzoate and N,N-dimethylbenzamide are metallated at the methyl group to form dipole-stabilized carbanions.[15] The organolithium intermediates produced in these reactions condense with the remaining ester or amide to afford aryl ketones of various types. Isocyanides[16] and dibromomethane[17] are selectively lithiated at the α-position by lithium 2,2,6,6-tetramethylpiperidide in the presence of carbonyl compounds, with which the organolithium compound subsequently reacts. The enolate anions of esters[5,18] and dianions of β-keto esters[19] and propiolic acid[20] have been formed by reaction with lithium 2,2,6,6-tetramethylpiperidide. Other reactions in which this hindered base has proven effective include the conversion of an epoxide to an enolate anion,[21] the generation of certain α-lithioorganoboranes,[22] the preparation of the highly strained tetracyclo[4.2.0.0^{2,4}.0^{3,5}]oct-7-ene from the appropriate tosylhydrazone,[23] and the insertion of magnesium into bacteriopheophytin α.[24] In many of these reactions, other bases, including less hindered amide bases such as lithium diisopropylamide, gave either lower yields or different products entirely.

1. Department of Chemistry, Herbert H. Lehman College, City University of New York, Bronx, New York 10468.
2. Department of Chemistry, The Pennsylvania State University, University Park, Pennsylvania 16802.
3. S. C. Watson and J. F. Eastham, *J. Organometal. Chem.*, **9**, 165 (1967).
4. R. A. Olofson and C. M. Dougherty, *J. Amer. Chem. Soc.*, **95**, 581 (1973).
5. R. A. Olofson and C. M. Dougherty, *J. Amer. Chem. Soc.*, **95**, 582 (1973).
6. W. K. Kirmse and H. Schütte, *Chem. Ber.*, **101**, 1674 (1968).
7. R. A. Olofson, K. D. Lotts, and G. N. Barber, *Tetrahedron Lett.*, 3779 (1976).
8. G. N. Barber and R. A. Olofson, *Tetrahedron Lett.*, 3783 (1976).
9. R. A. Olofson, K. D. Lotts, and G. N. Barber, *Tetrahedron Lett.*, 3381 (1976).
10. K. D. Lotts, Ph.D. Thesis, The Pennsylvania State University, 1976.
11. K. L. Shepard, *Tetrahedron Lett.*, 3371 (1975).
12. I. Fleming and T. Mah, *J. Chem. Soc., Perkin Trans. I*, 964 (1975).
13. I. Fleming and T. Mah, *J. Chem. Soc., Perkin Trans. I*, 1577 (1976).
14. C. J. Upton and P. Beak, *J. Org. Chem.*, **40**, 1094 (1975).
15. P. Beak and R. Farney, *J. Amer. Chem. Soc.*, **95**, 4771 (1973); P. Beak, B. G. McKinnie, and D. B. Reitz, *Tetrahedron Lett.*, 1839 (1977).
16. U. Schöllkopf, F. Gerhart, I. Hoppe, R. Harms, K. Hantke, K.-H. Scheunemann, E. Eilers, and E. Blume, *Justus Liebigs Ann. Chem.*, 183 (1976).
17. H. Taguchi, H. Yamamoto, and H. Nozaki, *Tetrahedron Lett.*, 2617 (1976).
18. S. R. Wilson and R. S. Meyers, *J. Org. Chem.*, **40**, 3309 (1975).
19. See also, S. N. Huckin and L. Weiler, *J. Amer. Chem. Soc.*, **96**, 1082 (1974).
20. B. S. Pitzele, J. S. Baran, and D. H. Steinman, *J. Org. Chem.*, **40**, 269 (1975).
21. L. S. Trzupek, T. L. Newirth, E. G. Kelley, N. E. Sbarbati, and G. M. Whitesides, *J. Amer. Chem. Soc.*, **95**, 8118 (1973).
22. M. W. Rathke and R. Kow, *J. Amer. Chem. Soc.*, **94**, 6854 (1972); R. Kow and M. W. Rathke, *J. Amer. Chem. Soc.*, **95**, 2715 (1973); D. S. Matteson and R. J. Moody, *J. Amer. Chem. Soc.*, **99**, 3196 (1977).
23. G. E. Gream, L. R. Smith, and J. Meinwald, *J. Org. Chem.*, **39**, 3461 (1974).
24. M. R. Wasielewski, *Tetrahedron Lett.*, 1373 (1977).

Appendix
Chemical Abstracts Nomenclature (Collective Index Number; Registry Numbers)

Lithium 2,2,6,6-tetramethylpiperidide: Piperidine, 2,2,6,6-tetra-methyl-, lithium salt (8,9); (38227-87-1)

1-Ethoxy-2-*p*-tolylcyclopropane: Ether, ethyl 2-*p*-tolylcyclopropyl (8); Benzene, 1-(2-ethoxycyclopropyl)-4-methyl- (9); *cis*-(40237-67-0); *trans*-(40489-59-6)

p-Xylene, α-chloro- (8); Benzene, 1-(chloromethyl)-4-methyl- (9); (104-82-5)

Ether, ethyl vinyl (8); Ethene, ethoxy- (9): (109-92-2)

Piperidine, 2,2,6,6-tetramethyl- (8,9); (768-66-1)

Citric acid (8); 1,2,3-Propanetricarboxylic acid, 2-hydroxy- (9); (77-92-9)

Hydrazine (8,9); (302-01-2)

4-Piperidone, 2,2,6,6-tetramethyl- (8); 4-Piperidinone, 2,2,6,6-tetramethyl- (9); (826-36-8)

Methyl bromide: Methane, bromo- (8,9); (74-83-9)

1,10-Phenanthroline (8,9); (66-71-7)

Lithium, butyl- (8,9); (109-72-8)

1-Methoxy-2-phenylcyclopropane: Ether, methyl 2-phenylcyclopropyl, trans- (8); Benzene, (2-methoxycyclopropyl)-, trans- (9); (26269-57-8)

Ether, dichloromethyl methyl (8); Methane, dichloromethoxy- (9); (4885-02-3)

Styrene (8); Benzene, ethenyl- (9); (100-42-5)

Methyl thiobenzoate: Benzoic acid, thio-, S-methyl ester (8); Benzenecarbothioic acid, S-methyl ester (9); (5925-68-8)

Benzamide, N,N-dimethyl- (8,9); (611-74-5)

Propiolic acid (8); 2-Propynoic acid (9); (471-25-0)

Tetracyclo[4.2.0.0^{2,4}.0^{3,5}]oct-7-ene (8,9); (35434-65-2)

Lithium diisopropylamide: Diisopropylamine, lithium salt (8); 2-Propanamine, N-(1-methylethyl)-, lithium salt (9); (4111-54-0)

Lithium, methyl- (8,9); (917-54-4)

CATALYTIC OSMIUM TETROXIDE OXIDATION OF OLEFINS: cis-1,2-CYCLOHEXANEDIOL

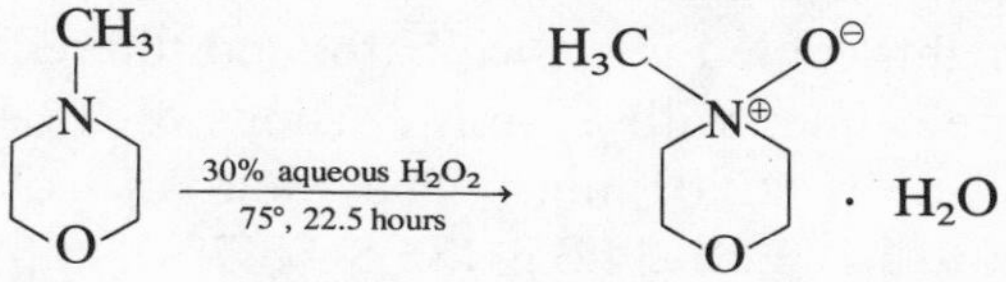

1

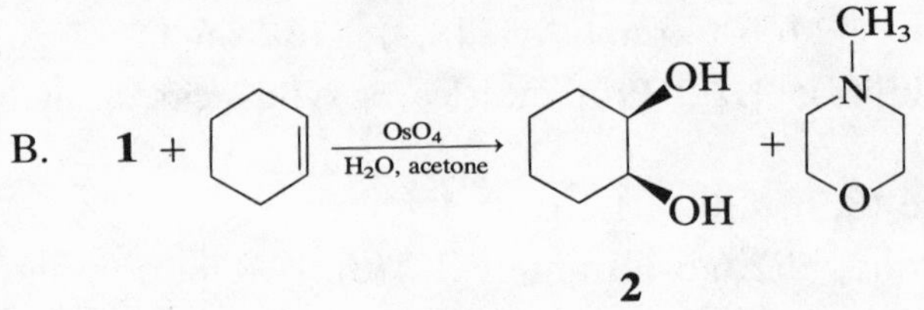

B.   **1** + (cyclohexene) $\xrightarrow[\text{H}_2\text{O, acetone}]{\text{OsO}_4}$ (cyclohexane-1,2-diol with OH, OH) + (N-methylmorpholine with CH₃, N, O)

2

Submitted by V. VanRheenen, D. Y. Cha, and W. M. Hartley[1]
Checked by N. Meyer, W. Wykypiel, and D. Seebach

1. Procedure

Caution! Care should be taken in handling osmium tetroxide. The vapor is toxic, causing damage to eyes, respiratory tract, and skin.

A. *N-Methylmorpholine N-oxide* (**1**) (Note 1). A 100-ml., three-necked, round-bottomed flask is equipped with a reflux condenser, a magnetic stirring bar, and a dropping funnel. The flask is flushed with nitrogen or argon and charged with 35.1 ml. (32.3 g., 0.32 mole) of N-methylmorpholine (Note 2). The flask is immersed in an oil bath maintained at 75°, and 29.1 ml. (32.4 g., 0.286 mole) of 30% aqueous hydrogen peroxide is added dropwise over a period of 2.5 hours (Note 3). The mixture is then stirred for 20 hours at 75°. At this time a negative peroxide test (KI paper) is obtained (Note 4). The reaction mixture is cooled to 50°, and a slurry of 50 ml. of methanol, 0.5 g. of charcoal, and 0.5 g. of Celite (Note 5) is added. After being stirred for 1 hour, the mixture is filtered and the filter cake washed three times with 15-ml. portions of methanol. The filtrate and combined washings are concentrated in a rotary evaporator (water aspirator vacuum), with the bath temperature finally reaching 95°, where it is held for 10 minutes. The flask is fitted with a reflux condenser, and the residual viscous oil is dissolved in 25 ml. of acetone at 60°. On cooling to 40° (with seeding, if crystals of the N-oxide are available) the product spontaneously crystallizes. The slurry is stored at room temperature overnight, cooled in an ice bath, and filtered. The crystals are washed three times with 15-ml. portions of 0° acetone and dried overnight at 40° (0.01 mm.) The yield of colorless crystalline monohydrate **1** is 32.4–34.3 g. (83.8–88.7%), m.p. 75–76° (Notes 6 and 7).

B. cis-1,2-*Cyclohexanediol*   (**2**).   A   250-ml.,   three-necked, round-bottomed flask, with a magnetic stirrer and a nitrogen inlet, is charged with 14.81 g. of monohydrate **1** (0.106 mole), 40 ml. of water, and 20 ml. of acetone. To this solution is added *ca.* 70 mg. of osmium tetroxide (0.27 mmole) (Note 8) and 10.1 ml. (8.2 g., 0.1 mole) of cyclohexene (Note 9). This two-phase solution is vigorously stirred under nitrogen at room temperature. The reaction is slightly exothermic and is maintained at room temperature with a water bath. During the overnight stirring period, the reaction mixture becomes homogeneous and light brown in color. After 18 hours, thin-layer chromatography (Note 10) shows the reaction to be complete. Sodium hydrosulfite (0.5 g.) (Note 11) and 5 g. of Magnesol (Note 12) slurried in 20 ml. of water are added, the slurry is stirred for 10 minutes, and the mixture is filtered through a pad of 5 g. of Celite on a 150-ml. sintered-glass funnel. The Celite cake is washed with three 15-ml. portions of acetone. The filtrate, combined with acetone wash, is neutralized to pH 7 with 6.4 ml. of 12*N* sulfuric acid. The acetone is evaporated under vacuum using a rotary evaporator. The pH of the resulting aqueous solution is adjusted to pH 2 with 2.3 ml. of 12*N* sulfuric acid, and the *cis*-diol **2** is separated from *N*-methylmorpholine hydrosulfate by extracting five times with 45-ml. portions of butanol (Note 13). The combined butanol extracts are extracted once with 25 ml. of 25% sodium chloride solution, and the aqueous phase is backwashed with 50 ml. of butanol. These butanol extracts are evaporated under vacuum to give 12.1 g. of white solid. The *cis*-diol **2** is separated from a small amount of insoluble solid (*ca.* 0.7 g.) by boiling the 12.1 g. of solid with a 200-ml., an 80-ml., and a 20-ml. portion of isopropyl ether (Note 14), decanting the solvent each time. The combined isopropyl ether fractions are evaporated to *ca.* 50 ml. under vacuum, and crystalline white plates precipitate. The mixture is cooled to *ca.* −15°. The crystals are filtered, washed with two 10-ml. portions of cold isopropyl ether, and dried to yield 10.18–10.32 g. (89–90%) *cis*-diol **2** (m.p. 96–97°).

2. Notes

1. *N*-Methylmorpholine *N*-oxide (**1**) can also be purchased from Eastman Organic Chemicals or Fluka A G.

2. Commercial material was used without purification (the purity was checked by refractive index and proton magnetic resonance).

3. The slow addition (2.5 hours) is important to avoid overheating of the reaction mixture. The potential danger of using hydrogen peroxide at an elevated temperature is minimized by using the 10% excess of N-methylmorpholine and by choosing reaction conditions that ensure rapid consumption, which avoids accumulation of peroxide in the mixture. A 50% aqueous hydrogen peroxide solution can also be used. The content of the commercial hydrogen peroxide (ca. 30 or 50%) must be determined by iodometric titration.

4. The very sensitive ether peroxide test strips (Merckoquant, Art. No. 10011), available from E. Merck, Darmstadt, are used. If the test is still positive at this point, an additional 0.2 ml. of N-methylmorpholine is added. Stirring and heating at 75° are continued for another 5 hours. Remaining peroxide renders the work-up and drying of the product potentially hazardous. N-Methylmorpholine N-oxide (1) and hydrogen peroxide form a strong 1:1 complex. In the reaction with osmium tetroxide, this complex produces conditions similar to those of the Milas reaction,[7] and some ketol formation may result.

5. Darco G 60, Aktivkohle, Fluka A G No. 05100, and Celite 512 Hyflosuper, Firma Schneider, Winterthur, Switzerland, No. 5100025, were used.

6. The procedure is designed to maintain the proper amount of water in the crystallization mixture so that the monohydrate 1 is obtained. It has the highest melting point and is the least hygroscopic. Other hydrated forms, such as the dihydrate (m.p. 35–60°) and mixed hydrates, may be isolated. A Karl Fischer assay of the water content is not necessary if the obtained material melts within the range given.

7. A second crop of 1.7–4.3 g. (4.3–11.1%) of the product can be obtained by evaporating the mother liquid, heating the residue at 97° for 20 minutes under reduced pressure, dissolving it in 55 ml. of acetone at 60°, and continuing as described.

8. Commercial osmium tetroxide was used without purification. It is not easy to accurately weigh this material because it rapidly sublimes.

9. Commercial cyclohexene was used (the purity was checked by refractive index). Addition of cyclohexene caused a darkening of the reaction mixture. This is caused by a finite concentration of the osmate ester. The reaction becomes lighter in color when complete.

10. The reaction may be followed by thin-layer chromatography. The ratio of the R_f values for cyclohexene and cis-diol 2 is 2 : 1 (commercial silica gel plates, ethyl acetate). The plates are best visualized by first spraying with 1% aqueous potassium permanganate, then with methanolic sulfuric acid, followed by charring with heat. The checkers found that, if the procedure is followed exactly, monitoring the reaction by thin-layer chromatography is unnecessary.

11. Sodium hydrosulfite reduces the osmium tetroxide to insoluble lower-valent osmium species.

12. The submitter used Magnesol, industrial grade, available from Reagent Chemical Research, Inc., Pilot Engineering Division. The checkers used Florisil TLC, available from E. Merck, Darmstadt, No. 12519.

13. Since the cis-diol 2 is very water soluble, a polar solvent such as butanol is required to extract it. Butanol forms an efficient water azeotrope. More conventional solvents may be used for less polar products.

14. Isopropyl ether readily forms explosive peroxides. It should be tested for peroxides, and contact with air should be minimized.

3. Discussion

cis-Dihydroxylation of olefins may be effected with potassium permanganate, osmium tetroxide, or silver iodoacetate according to Woodward's procedure.[2] Oxidation of cyclohexene to cis-diol 2 with potassium permanganate is reported to proceed in only 30–40% yields.[3,4] A modification of Woodward's procedure, in which iodine, potassium iodate and potassium acetate in acetic acid were used, has given cis-diol 2 in 86% yield.[5] This procedure is particularly useful for placement of cis-diols on the more hindered side of more complex substrates.

The reaction of an olefin with osmium tetroxide is the most reliable method for cis-dihydroxylation of a double bond, particu-

TABLE I
Preparation of *cis*-Diols by Catalytic Oxidation of Olefins with Osmium Tetroxide

Starting Material	Product	Procedure	(Isolated yields, %)	Reference
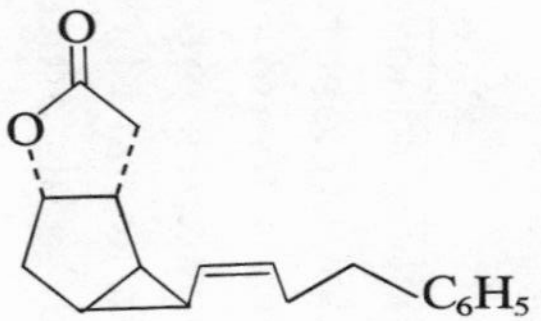	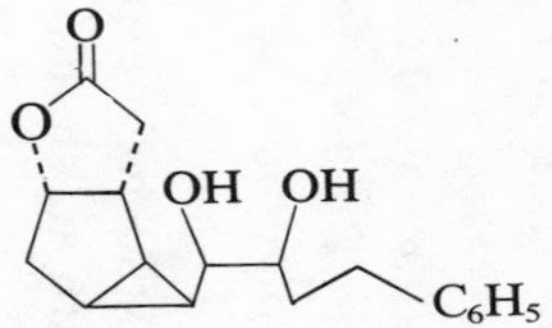	NMO–OsO$_4$	(>95)	18
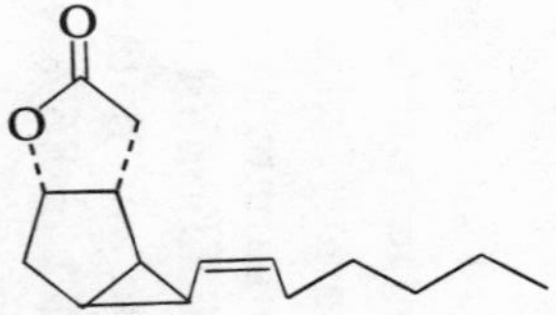	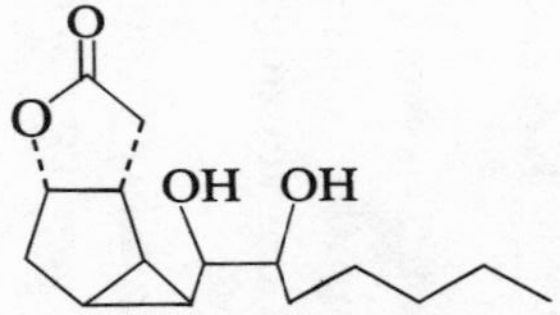	NMO–OsO$_4$	(>95)	10[a]
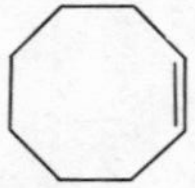	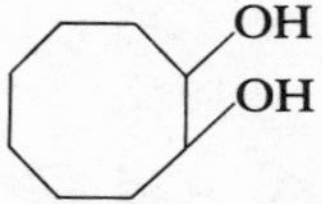	NMO–OsO$_4$ NaClO$_3$–OsO$_4$ KMnO$_4$ H$_2$O$_2$–OsO$_4$	(79) (30) (50) (11.4)	10[a] 11 12 11
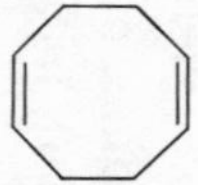	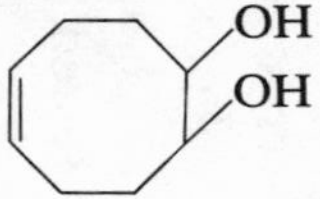	NMO–OsO$_4$ OsO$_4$, 1 mole KMnO$_4$	(31) (14) (3)	10[a] 13 14

48

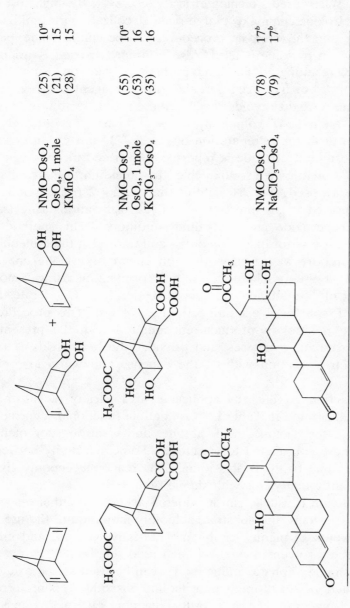

NMO–OsO$_4$	(25)	10[a]
OsO$_4$, 1 mole	(21)	15
KMnO$_4$	(28)	15
NMO–OsO$_4$	(55)	10[a]
OsO$_4$, 1 mole	(53)	16
KClO$_3$–OsO$_4$	(35)	16
NMO–OsO$_4$	(78)	17[b]
NaClO$_3$–OsO$_4$	(79)	17[b]

[a] The reaction was carried out in aqueous acetone at room temperature using 0.2–1.0 mole % OsO$_4$ (see Experimental section).

[b] Solvent composition of 10:3:1 tert-butanol–tetrahydrofuran–water was preferred for this reaction.

49

larly for preparation of *cis*-diols on the least hindered side of the molecule. When used stoichiometrically, however, the high cost of osmium tetroxide can make a large-scale glycolization prohibitively expensive, and the work-up procedures can be cumbersome, particularly when pyridine is used. Also osmium tetroxide is volatile and toxic, resulting in handling problems. Catalytic osmylation using chlorate[6] or hydrogen peroxide (Milas reagent[7]) to regenerate osmium tetroxide avoids some of these problems, but overoxidation to an α-ketol commonly leads to losses in yield and to separation problems. Preparation of *cis*-diol **2** with sodium chlorate and osmium tetroxide is reported to proceed in 46% yield,[3] and in 76% yield[8] when sodium chlorate, potassium osmate, and a detergent are used. A 62% yield of the *cis*-diol **2** from cyclohexene is reported in an interesting catalytic osmylation using *tert*-butyl hydroperoxide under alkaline conditions.[9] This method is particularly useful for oxidation of tri- and tetrasubstituted olefins.

In this report we describe the conversion of cyclohexene to *cis*-diol **2** in 90% yield in a catalytic osmylation using 1 mole equivalent of *N*-methylmorpholine *N*-oxide (**1**) to regenerate the less than 1 mole % of osmium tetroxide catalyst. This procedure avoids the α-ketol by-products encountered with the presently available catalytic processes, and provides the high yields of the stoichiometric reaction without the expense and work-up problems.

The reaction is generally applicable to a variety of substrate types, as illustrated in Table I.[10] Compatible functionality includes hydroxyl, ester, lactone, acid, ketone, and electron-poor olefins such as those conjugated to α-ketones. Some selectivity between isolated double bonds is also found. The reaction generally gives nearly quantitative yields with simple olefins.

The reaction is usually run in aqueous acetone in either one- or two-phase systems, but substrate solubility may require the use of other solvents. Aqueous *tert*-butanol, tetrahydrofuran, and mixtures of these solvents have also been used successfully.

Other simple aliphatic amine oxides can be used as the oxidant in this reaction, but *N*-methylmorpholine *N*-oxide (**1**) is preferred because it generally gives a faster reaction rate and is easily prepared. The reaction can also be used to convert aliphatic amine oxides into amines.

1. The Upjohn Company, Kalamazoo, Michigan 49001.
2. F. D. Gunstone, in "*Advances in Organic Chemistry*," **17**, Vol. 1, edited by R. A. Raphael, E. C. Taylor, and H. Wynberg, Interscience Publishers, New York, N. Y., 1960, p. 110ff.
3. M. F. Clark and L. N. Owen, *J. Chem. Soc.* (Lond.), 315 (1949).
4. K. B. Wiberg and K. A. Saegebarth, *J. Amer. Chem. Soc.*, **79**, 2822 (1957).
5. L. Mangoni, M. Adinolfi, G. Barone, and M. Parrilli, *Tetrahedron Lett.*, 4485 (1973).
6. K. A. Hofmann, *Chem. Ber.*, **45**, 3329 (1912).
7. N. A. Milas and S. Sussman, *J. Amer. Chem. Soc.*, **58**, 1302 (1936); N. A. Milas, J. H. Trepagnier, J. T. Nolan, Jr., and M. I. Iliopulos, *J. Amer. Chem. Soc.*, **81**, 4730 (1959); C. J. Norton and R. E. White, "Selective Oxidation Processes," *Advances in Chemistry Series*, No. 51, American Chemical Society, Washington, D. C., 1965, pp. 10–25.
8. W. D. Lloyd, B. J. Navarette, and M. F. Shaw, *Synthesis*, 610 (1972).
9. K. B. Sharpless and K. Akashi, *J. Amer. Chem. Soc.*, **98**, 1986 (1976).
10. V. VanRheenen, R. C. Kelly, and D. F. Cha, *Tetrahedron Lett.*, 1973 (1976).
11. A. C. Cope, S. W. Fenton, and C. F. Spencer, *J. Amer. Chem. Soc.*, **74**, 5884 (1952).
12. W. P. Weber and J. P. Shepherd, *Tetrahedron Lett.*, 4907 (1972).
13. K. Tanaka, *J. Biol. Chem.*, **247**, 7465 (1972).
14. J. L. Jernow, D. Gray, and W. D. Clossen, *J. Org. Chem.*, **36**, 3511 (1971).
15. Y. F. Shealy and J. D. Clayton, *J. Amer. Chem. Soc.*, **91**, 3075 (1969).
16. R. C. Kelly and I. Schletter, *J. Amer. Chem. Soc.*, **95**, 7156 (1973).
17. W. P. Schneider and A. V. McIntosh, U.S. Patent 2,769,824 (1957) [*C.A.* **51**, 8822e (1957)]. The use of NMO in catalytic OsO_4 reactions was first disclosed in this patent during work to introduce the corticoid side chain (an α-ketol) in a steroid.
18. B. J. Magerlein, G. L. Bundy, F. H. Lincoln, and G. A. Youngdale, *Prostaglandins*, **9**(1), 5 (1975).

Appendix
Chemical Abstracts Nomenclature (Collective Index Number; Registry Numbers)

Osmium tetroxide: Osmium oxide (OsO_4) (8,9); (20816-12-0)

1,2-Cyclohexanediol, *cis*- (8,9); (1792-81-0)

N-Methylmorpholine *N*-oxide: Morpholine, 4-methyl-, 4-oxide (8,9); (7529-22-8)

N-Methylmorpholine: Morpholine, 4-methyl- (8,9); (109-02-4)

Hydrogen peroxide (8,9); (7722-84-1)

N-Methylmorpholine *N*-oxide monohydrate: Morpholine, 4-methyl-, 4-oxide, monohydrate (8,9); (−)

Cyclohexene (8,9); (110-83-8)

Sodium hydrosulfite: Dithionous acid, disodium salt (8,9); (7775-14-6)

Magnesol: Silicic acid, magnesium salt (8,9); (1343-88-0)

N-Methylmorpholine hydrosulfate: Morpholine, 4-methyl-, hydrosulfate (8,9); (−)

Sodium chloride (8,9); (7627-14-5)

Isopropyl ether (8); Propane, 2,2'-oxybis- (9); (108-20-3)

Morpholine (8,9); (110-91-8)

Potassium permanganate: Permanganic acid ($HMnO_4$), potassium salt (8,9); (7722-64-7)

tert-Butyl hydroperoxide (8); Hydroperoxide, 1,1-dimethylethyl- (9); (75-91-2)

COPPER CATALYZED ARYLATION OF β-DICARBONYL COMPOUNDS: 2-(1-ACETYL-2-OXOPROPYL)BENZOIC ACID

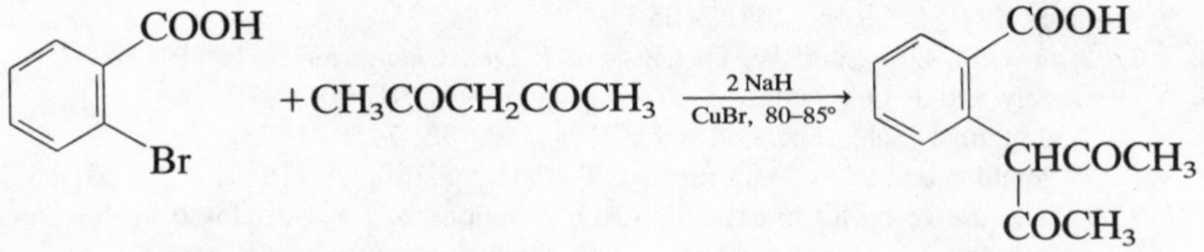

Submitted by ALLE BRUGGINK, STEPHEN J. RAY, and ALEXANDER McKILLOP[1]
Checked by SADAO HAYASHI and WATARU NAGATA

1. Procedure

A 250-ml., three-necked, round-bottomed flask is equipped with a sealed-mechanical stirrer, a Claisen adapter fitted with a thermometer and a gas-inlet adapter, and a drying tube (Note 1) filled with calcium chloride. The flask is charged with 150 ml. of acetylacetone (Notes 2 and 3), 25 g. (0.125 mole) of 2-bromobenzoic acid (Note 4), and 1.0 g. of copper(I) bromide (Note 5). The mixture is thoroughly purged with dry nitrogen and stirred rapidly while a total of 9.0 g. (0.3 mole) of an 80% dispersion of sodium hydride in mineral oil (Note 6) is slowly added portionwise through the inlet protected by the calcium chloride drying tube. Addition of the first portions of the sodium hydride results in an immediate exothermic reaction, and the temperature of the mixture rises rapidly to 50–55°. The remainder of the sodium hydride is added at such a rate that the temperature remains within the 50–55° range, and external cooling with a cold water bath may be necessary (Note 7). After addition of the sodium hydride is complete (30–35 minutes), the flask is placed in a hot-water bath heated to

80–85°, and the reaction mixture is stirred and heated for 5 hours, during which time a slow stream of dry nitrogen is passed through the equipment.

When the reaction mixture has cooled, the contents of the flask are poured into a 1-l. Erlenmeyer flask containing 400 ml. of distilled water. The reaction flask and stirrer are thoroughly washed with an additional 100 ml. of distilled water, the washings are added to the bulk of the reaction mixture, and the aqueous mixture is allowed to stand at room temperature for 15 minutes to ensure completion of hydrolysis and precipitation of inorganic salts. The salts are removed by filtration under reduced pressure (Note 8) and discarded. The filtrate is transferred to a 1-l. separatory funnel, and the excess acetylacetone is separated (Note 9). The aqueous phase is washed with five 100-ml. portions of ethyl ether (Note 9) and transferred to a 1-l. conical flask. Nitrogen is blown through the solution for 15 minutes to remove traces of ethyl ether (Note 10). The aqueous solution is then acidified to pH 3 with concentrated hydrochloric acid, the flask being constantly swirled during addition of the acid, and the mixture is allowed to stand at room temperature for 30 minutes to ensure complete precipitation of the product. The crude material is collected by filtration under reduced pressure, washed with 25 ml. of distilled water, and dried under reduced pressure over phosphorus pentoxide to give 21–22 g. (76–80%) of crude product, m.p. 138–145° (Note 11). Recrystallization from a mixture of 50 ml. of methanol and 100 ml. of water gives 19.5–21 g. (71–76%) of pure material (Note 12) as colorless prisms, m.p. 142–144.5° (Note 11).

2. Notes

1. The checkers inserted a condenser between the flask and the drying tube to prevent acetylacetone from being carried away by the nitrogen stream.

2. The acetylacetone was washed with aqueous sodium bicarbonate solution, dried over sodium sulfate, and distilled prior to use. The checkers used acetylacetone of higher than 95% purity purchased from Wako Pure Chemical Industries, Ltd., Japan, and purified by distillation prior to use.

3. The use of a large amount of acetylacetone as both reagent

and solvent is essential to prevent precipitation of the sodium salts of acetylacetone and 2-bromobenzoic acid. If this happens, the whole reaction mixture rapidly solidifies, resulting in incomplete mixing and in poor and irreproducible yields of the product.

4. Commercial 2-bromobenzoic acid, purchased from Aldrich Chemical Company, Inc., is a gray powder, m.p. 144–147°, and was purified as follows: the crude acid was dissolved in warm $2N$ sodium hydroxide solution; the mixture was heated to reflux, treated with activated carbon, filtered, and cooled, and the filtrate was acidified with concentrated hydrochloric acid. The colorless solid that precipitated was collected by filtration under reduced pressure and recrystallized from aqueous methanol to give pure 2-bromobenzoic acid as colorless needles, m.p. 148–150°. The checkers used 2-bromobenzoic acid of m.p. 151–151.5°, obtained from Wako Pure Chemical Industries, Ltd., Japan, without further purification. Their results were comparable to those of the submitters.

5. Copper(I) bromide was prepared according to J. L. Hartwell, *Org. Syn.*, Coll. Vol. **3,** 186 (1955); the salt was dried under reduced pressure over P_2O_5 prior to use. The checkers observed that good results could be realized by using a commercially available copper(I) bromide of higher than 95% purity from Wako Pure Chemical Industries, Ltd., Japan, without purification.

6. The checkers used a 50% dispersion of sodium hydride in mineral oil, available from Wako Pure Chemical Industries, Ltd., Japan, and obtained results comparable to those of the submitters.

7. Cooling is normally required. It is important, however, to ensure that the temperature of the reaction mixture does not fall below 50°; otherwise efficient stirring becomes impossible.

8. The inorganic salts that precipitate at this stage are generally in a very finely divided form. A medium-sized Büchner funnel (11–15 cm.) or a small Büchner funnel (5–7.5 cm.) fitted with a Celite pad should be used to avoid slow filtration.

9. The acetylacetone can readily be recovered and recycled. The bulk of it, which is obtained at the separation step, is combined with the ether washings, and the solution is dried over anhydrous sodium sulfate. The solvent is then removed by evaporation under reduced pressure, and the residual crude acetylacetone is purified by distillation to give a recovery of 85–95 g. of pure acetylacetone.

10. Acidification without prior removal of the small amount of ethyl ether present in the aqueous solution results in precipitation of the product as an oily, semisolid mass that is difficult to filter.

11. Evolution of water is noticeable from *ca.* 130° upward; this is a result of lactone formation between the enolic hydroxyl group of the β-dicarbonyl unit and the aromatic carboxylic acid group.

12. The product, which exists entirely in the enolic form, has the following spectral data: ultraviolet (methanol) nm. max. (ε): 227 (7760) and 287 (8190); infrared (Nujol) cm^{-1}: 3300–2400, 1700–1690, 1305, 1270, 810, 770, 720, and 700; proton magnetic resonance (chloroform-*d*) δ (multiplicity, number of protons, assignment): 1.82 (sharp singlet, 6, two CH_3), 7.2–8.2 (complex multiplet, 4, C$_6H_4$), 10.1 (broad singlet, 1, COOH), and 16.3 (broad singlet, 1, enolic OH).

3. Discussion

2-(1-Acetyl-2-oxopropyl)benzoic acid has been prepared by the copper-catalyzed arylation of acetylacetone with 2-bromobenzoic acid. Facile condensation of β-dicarbonyl compounds with 2-bromobenzoic acid was first demonstrated by Hurtley in 1929,[2] and reactions of this type have occasionally been employed with limited success in a number of natural product syntheses.[3] The reaction conditions originally employed for these condensations and subsequently adopted by all other workers consist in the use of sodium ethoxide as base, ethanol as solvent, and copper powder as catalyst. Under these conditions, however, 2-ethoxybenzoic acid is obtained as a by-product in substantial amounts (25–35%), together with smaller amounts of unchanged 2-bromobenzoic acid (5–10%), and separation and purification of the desired α-(2-carboxyphenyl)-β-dicarbonyl product is often difficult and tedious.

The present method of preparation is that described by Bruggink and McKillop.[4] It has the particular advantages of high yield and manipulative simplicity, and avoids the problem inherent in Hurtley's procedure of separation of mixtures of carboxylic acids by fractional crystallization or column chromatography. The method is, moreover, of wide applicability with respect to both the β-dicarbonyl compound and the 2-bromobenzoic acid. The synthetic scope and limitations of this procedure for the direct arylation of

β-dicarbonyl compounds have been fully defined with respect to a wide range of substituted 2-bromobenzoic acids and to certain other bromoaromatic and heteroaromatic carboxylic acids.[4]

1. School of Chemical Sciences, University of East Anglia, Norwich, England.
2. W. R. H. Hurtley, *J. Chem. Soc.*, 1870 (1929).
3. K. A. Cirigottis, E. Ritchie, and W. C. Taylor, *Aust. J. Chem.*, **27**, 2209 (1974).
4. A. Bruggink and A. McKillop, *Tetrahedron*, **31**, 2607 (1975).

Appendix
Chemical Abstracts Nomenclature (Collective Index Number; Registry Numbers)

Benzoic acid, *o*-acetylacetonyl- (8); Benzoic acid, 2-(1-acetyl-2-oxopropyl)- (9); (52962-26-2)

Acetylacetone: 2,4-Pentanedione (8,9); (123-54-6)

Benzoic acid, *o*-bromo- (8); Benzoic acid, 2-bromo- (9); (88-65-3)

Copper(I) bromide: Copper bromide (CuBr) (8,9); (7787-70-4)

Sodium hydride (8,9); (7646-69-7)

Ethyl ether (8); Ethane, 1,1'-oxybis- (9); (60-29-7)

Benzoic acid, *o*-ethoxy- (8); Benzoic acid, 2-ethoxy- (9); (134-11-2)

CYCLOPENTENONES FROM α,α'-DIBROMOKETONES AND ENAMINES: 2,5-DIMETHYL-3-PHENYL-2-CYCLOPENTEN-1-ONE

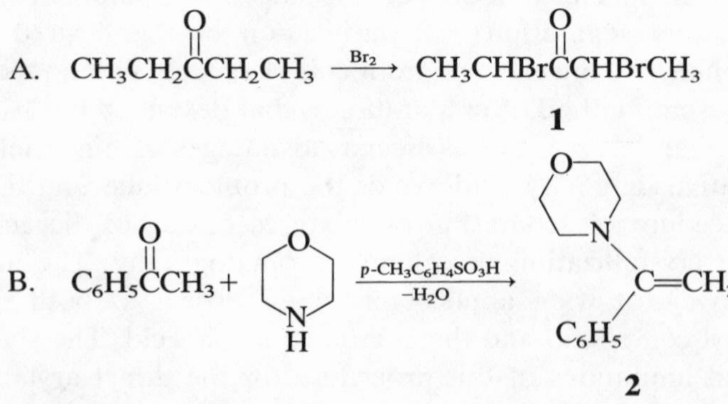

A. $CH_3CH_2\overset{\text{O}}{\overset{\|}{C}}CH_2CH_3 \xrightarrow{Br_2} CH_3CHBr\overset{\text{O}}{\overset{\|}{C}}CHBrCH_3$

1

B. $C_6H_5\overset{\text{O}}{\overset{\|}{C}}CH_3 +$ $\xrightarrow[\text{$-H_2O$}]{p\text{-}CH_3C_6H_4SO_3H}$

2

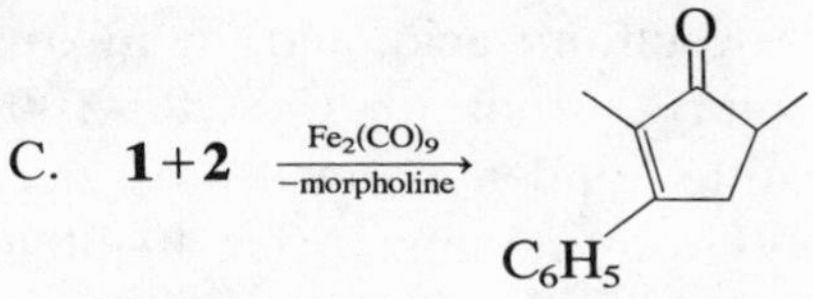

C. **1+2** $\xrightarrow[-\text{morpholine}]{\text{Fe}_2(\text{CO})_9}$

3

Submitted by R. Noyori, K. Yokoyama, and Y. Hayakawa[1]
Checked by Michael J. Haire and William A. Sheppard

1. Procedure

Caution! See benzene warning, p. 168. The reaction in Part C should be carried out in a well-ventilated hood because iron carbonyls are highly toxic.

A. *2,4-Dibromo-3-pentanone* (**1**). A 300-ml., three-necked, round-bottomed flask is equipped with a magnetic stirrer, a thermometer, and a 125-ml. pressure-equalizing dropping funnel, connected to a trap for absorbing hydrogen bromide evolved during the reaction (Note 1). The flask is charged with 43.0 g. (0.500 mole) of diethyl ketone and 100 ml. of 47% hydrobromic acid. The dropping funnel is charged with 160 g. (1.00 mole) of bromine. The bromine is added with stirring over a 1-hour period, and the temperature of the reaction mixture increases to 50–60°. After addition is complete, stirring is continued for an additional 10 minutes, and then 100 ml. of water is added to the reaction mixture. The separated heavy organic layer is washed with 30 ml. of saturated aqueous sodium bisulfite. The brownish organic solution is dried over calcium chloride and distilled under reduced pressure through a 15-cm. vacuum-jacketed Vigreux column to give 85.2–92.5 g. (70–76%) of 2,4-dibromo-3-pentanone (**1**) (Note 2), b.p. 51–57° (3 mm.), as a slightly yellow liquid.

B. *α-Morpholinostyrene* (**2**). A mixture of 75.0 g. (0.625 mole) of acetophenone, 81.0 g. (0.930 mole) of morpholine (Note 3), 200 mg. of p-toluenesulfonic acid, and 250 ml. of benzene is placed in a 500-ml., round-bottomed flask equipped with a water separator (Note 4), under a reflux condenser protected by a calcium chloride drying tube. Separation of water begins with reflux and is complete after 180 hours. After the mixture is cooled to room temperature, 200 mg. of sodium ethoxide is added for removal of p-toluenesulfonic acid, and the mixture is concentrated using a vacuum rotary evaporator (50°, 80–100 mm.). The crude oily product is distilled under reduced pressure through a 15-cm. vacuum-jacketed Vigreux column. After 40–50 ml. of a mixture of morpholine and acetophenone is recovered as a forerun at 40–90° (20 mm.), 67.5–75.4 g. (57–64%) of α-morpholinostyrene is collected as a pale yellow liquid, b.p. 85–90° (0.03 mm.) (Note 5).

C. 2,5-Dimethyl-3-phenyl-2-cyclopenten-1-one (3). A 1-l., three-necked, round-bottomed flask equipped with a sealed mechanical stirrer, a rubber septum, and a bubbler filled with liquid paraffin is charged with 40.0 g. (0.110 mole) of diiron nonacarbonyl (Note 6) and 250 ml. of dry benzene (Note 7). The system is flushed with nitrogen, and 56.8 g. (0.300 mole) of α-morpholinostyrene (Note 8) and 24.4 g. (0.100 mole) of 2,4-dibromo-3-pentanone (1) are injected by syringe through the rubber septum. The flask is immersed in a 32° bath, and the reaction mixture is stirred under a nitrogen atmosphere (Note 9). After 20 hours 230 g. of silica gel (Note 10) and 100 ml. of benzene are added. The resulting slurry is stirred at 32° for an additional 2.5 hours (Note 11). The whole mixture is poured onto a silica gel pad (Notes 10 and 12) with the aid of 200 ml. of ethyl ether, and the pad is washed with 1 l. of ethyl ether (Notes 13 and 14). The combined organic solutions are concentrated on a vacuum rotary evaporator (Note 13) to give 35–45 g. of a brown oil. This is the desired cyclopentenone 3, contaminated by acetophenone formed by decomposition of the excess enamine. The oil is distilled under reduced pressure by using a short-path distillation apparatus (Note 15). The forerun of 20–25 g., b.p. 35–50° (0.1 mm.), is recovered acetophenone. At 100–105° (0.02 mm.), 12.0–12.4 g. [64–67% yield (Note 14)] of cyclopentenone 3 is obtained as a pale yellow oil that crystallizes on cooling with ice water. An analytical sample was prepared by recrystallization from hexane as colorless needles, m.p. 57–59° (Note 16).

2. Notes

1. See Figure 7 in reference 2.
2. The submitter reported a yield of 116 g. (95%). Care should be taken to prevent the dibromoketone from coming into contact with the skin; allergic reactions have been observed in several

cases. Also, the checkers found the crude product to have lachrymatory properties. Immediate use after distillation is recommended if high yield is to be obtained in the next step.

3. An excess of morpholine is required because a considerable amount is lost with the water that separates during the reaction.

4. See Figure 12 in reference 3.

5. The distilled product is 97% pure and contaminated with 3% acetophenone (nuclear magnetic resonance analysis). Since the enamine is easily hydrolyzed and deteriorates on long standing, use of a freshly-distilled material is recommended. The checkers found that α-morpholinostyrene contaminated with 20% acetophenone could be used for the next step without any significant reduction in yield.

6. Diiron nonacarbonyl is available from Alpha Inorganics, Inc., or Strem Chemicals, Inc. The submitters made the complex through photolysis of iron pentacarbonyl by the method of King.[4] Procedures for preparation are also given by E. H. Braye and W. Hübel, *Inorg. Syn.*, **8,** 178 (1966), in which the name diiron enneacarbonyl is used.

7. The submitters used benzene distilled from lithium aluminum hydride, but the checkers used ACS-grade benzene as well as benzene distilled from lithium aluminum hydride, with no significant change in yield.

8. The submitters obtained a markedly lower yield of product when an excess of the enamine was not used.

9. Evolution of carbon monoxide begins in a few minutes after mixing the starting materials, continuing for *ca.* 3 hours. Cessation of gas evolution does not necessarily mean completion of the cyclocoupling reaction.

10. The submitters used Merck Kieselgel 60 (70–230 mesh ASTM).

11. This procedure is for elimination of morpholine from the labile primary product, 2,5-dimethyl-3-morpholino-3-phenyl-cyclopentanone.

12. The 150 g. of silica gel is packed in a 13 (diameter) × 12-cm. (length) glass filter.

13. The filter cake and the distillate must be treated with nitric acid to decompose the contaminates of iron carbonyl complexes. This treatment should be done very carefully in a well-ventilated hood, because carbon monoxide is evolved vigorously.

14. The submitters reported a yield of 14.8–15.6 g. (80–84%) based on the starting dibromide. The submitters report that the cyclopentenone product seems to absorb on silica gel, and 1 l. of ethyl ether is required to attain complete extraction. A smaller quantity of ether wash was used by the checkers. Proton magnetic resonance analysis of the crude mixture before distillation indicated the formation of the cyclopentenone 3 in 83–87% yield.

15. See Figure 2 in reference 5.

16. The spectral properties of pure product are as follows: infrared (carbon tetrachloride) cm^{-1}: 1696 (conjugated C=O) and 1626 (conjugated C=C); ultraviolet (ethanol solution) nm. max. (ε): 220 (5220) and 279 (11,200); proton magnetic resonance (carbon tetrachloride) δ (multiplicity, number of protons, assignment, coupling constant J in Hz.): 1.22 (doublet, 3, CHCH_3, $J = 7.0$), 1.91 (triplet, 3, vinyl CH_3, $J = 2.0$), 2.1–2.7 (multiplet, 2, CHCH$_3$ and a methylene proton cis to CH$_3$ group), 3.14 (doublet of doublet of quartet, 1, a methylene proton trans to CH$_3$ group, $J = 18$, 7.5, and 2.0), and 7.38 (multiplet, 5, aromatic H).

3. Discussion

The starting materials, 2,4-dibromo-3-pentanone[6] and α-morpholinostyrene,[7] have been prepared in satisfactory yields by known procedures after slight modifications. The procedure of the $3 + 2 \rightarrow 5$ cyclocoupling reaction is essentially that described originally by the submitters.[8] The main advantages of this procedure are the directness, the ready availability of the starting materials, and the wide generality for the preparation of 2,5-dialkyl-2-cyclopenten-1-ones. Several examples are given in Table I. Enamines derived from aldehydes, open-chain ketones, and cyclic ketones can be employed generally. The method has been extended to the synthesis of spiro[4.n]alkenones and certain azulene derivatives.[8] A reaction mechanism for the cyclocoupling reaction has been advanced.[9] The reactive oxyallyl intermediates generated from dibromoketones and iron carbonyls can be trapped efficiently by reagents such as enamines,[8] aromatic olefins,[10] 1,3-dienes,[11] furans,[12] carboxamides,[13] and N-carboalkoxypyrroles.[14] At present, dibromides of methyl ketones cannot be used as the starting material. However, polybromoketones, including $\alpha,\alpha,\alpha',\alpha'$-

tetrabromoacetone, serve as a precursor of the reactive species in certain cases, and the coupling reactions have been applied to various naturally occurring products.[15-18]

1. Department of Chemistry, Nagoya University, Chikusa, Nagoya 464, Japan.
2. C. S. Marvel and W. M. Sperry, Org. Syn., Coll. Vol. 1, 95 (1941).
3. S. Natelson and S. Gottfried, Org. Syn., Coll. Vol. 3, 381 (1955).
4. R. B. King, Organometal. Syn., 1, 93 (1965).
5. W. Nagata and M. Yoshioka, Org. Syn., 52, 90 (1972).
6. C. Rappe, Acta Chem. Scand., 16, 2467 (1962).
7. S. Hünig, K. Hübner, and E. Benzing, Chem. Ber., 95, 926 (1962).
8. R. Noyori, K. Yokoyama, S. Makino, and Y. Hayakawa, J. Amer. Chem. Soc., 94, 1772 (1972).
9. R. Noyori, Y. Hayakawa, M. Funakura, H. Takaya, S. Murai, R. Kobayashi, and S. Tsutsumi, J. Amer. Chem. Soc., 94, 7202 (1972).
10. R. Noyori, K. Yokoyama, and Y. Hayakawa, J. Amer. Chem. Soc., 95, 2722 (1973).
11. R. Noyori, S. Makino, and H. Takaya, J. Amer. Chem. Soc., 93, 1272 (1971).
12. R. Noyori, Y. Baba, S. Makino, and H. Takaya, Tetrahedron Lett., 1741 (1973).
13. R. Noyori, Y. Hayakawa, S. Makino, N. Hayakawa, and H. Takaya, J. Amer. Chem. Soc., 95, 4103 (1973).
14. R. Noyori, S. Makino, Y. Baba, and Y. Hayakawa, Tetrahedron Lett., 1049 (1974).
15. R. Noyori, Y. Baba, and Y. Hayakawa, J. Amer. Chem. Soc., 96, 3336 (1974).
16. R. Noyori, S. Makino, T. Okita, and Y. Hayakawa, J. Org. Chem., 40, 806 (1975).
17. Y. Hayakawa, M. Sakai, and R. Noyori, Chem. Lett., 509 (1975).
18. R. Noyori, T. Souchi, and Y. Hayakawa, J. Org. Chem., 40, 2681 (1975).

Appendix
Chemical Abstracts Nomenclature (Collective Index Number; Registry Numbers)

α,α,α',α'-Tetrabromoacetone: 2-Propanone, 1,1,3,3-tetrabromo- (8,9); (22612-89-1)

Acetophenone (8); Ethanone, 1-phenyl- (9); (98-86-2)

2-Cyclopenten-1-one, 2,5-dimethyl-3-phenyl- (8,9); (36461-43-5)

Morpholine (8,9); (110-91-8)

3-Pentanone, 2,4-dibromo- (8,9); (815-60-1)

α-Morpholinostyrene: Morpholine, 4-(1-phenylvinyl)- (8); Morpholine, 4-(1-phenylethenyl)- (9); (7196-01-2)

p-Toluenesulfonic acid (8); Benzenesulfonic acid, 4-methyl- (9); (104-15-4)

Diiron nonacarbonyl: Iron, tri-μ-carbonylhexacarbonyldi- [Fe–Fe] (8,9); (15321-51-4)

Iron pentacarbonyl (8); Iron carbonyl (Fe(CO)$_5$) (9); (13463-40-6)

TABLE I
Iron Carbonyl–Promoted Cyclopentenone Synthesis[a]

Dibromide	Enamine	Product[b]	Yield (%)[c]
$CH_3CHBrCCHBrCH_3$ (with C=O)	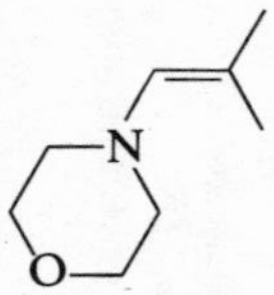	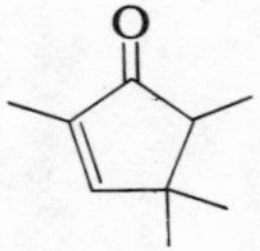	79
$(CH_3)_2CHCHBrCCHBrCH(CH_3)_2$ (with C=O)	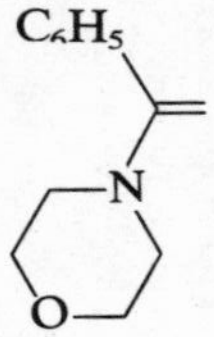	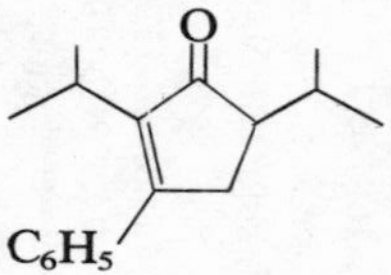 C_6H_5	72
$CH_3CHBrCCHBrCH_3$ (with C=O)	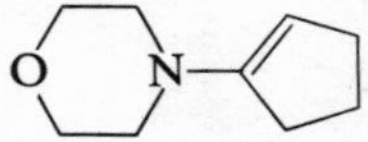	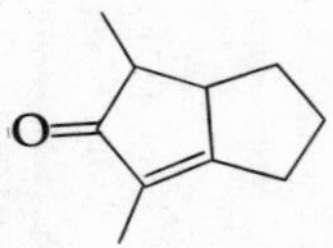	74
$CH_3CHBrCCHBrCH_3$ (with C=O)	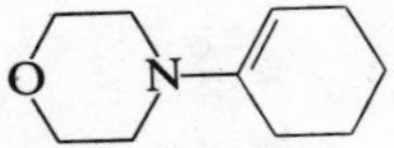	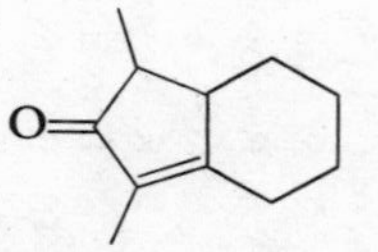	100

(CH$_3$)$_2$CHCHCHBrCCHBrCH(CH$_3$)$_2$ with carbonyl

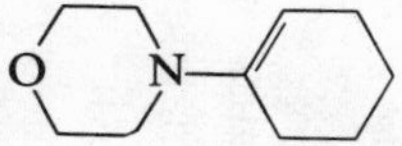

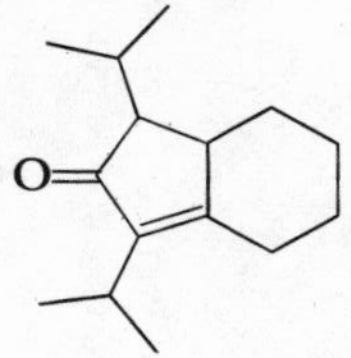

CH$_3$CHBrCCHBrCH$_3$ with carbonyl

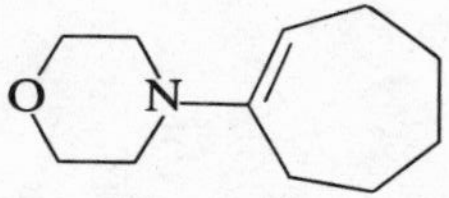

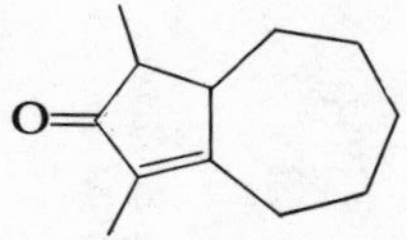

CH$_3$CHBrCCHBrCH$_3$ with carbonyl

73

100

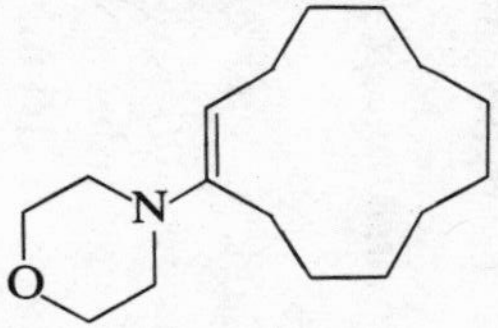

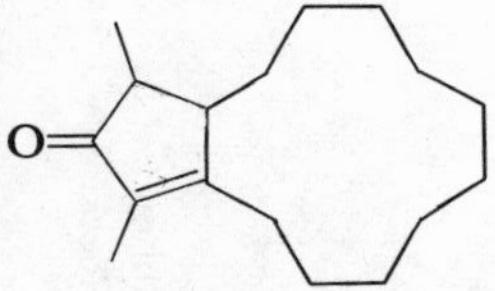

90

[a] Reference 8.
[b] A mixture of epimers, when possible, is obtained.
[c] Isolated yield based on starting dibromide.

63

cis-DICHLOROALKANES FROM EPOXIDES:
cis-1,2-DICHLOROCYCLOHEXANE

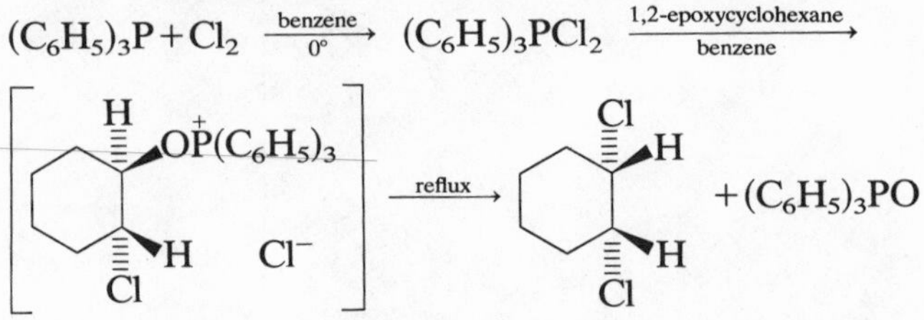

$$(C_6H_5)_3P + Cl_2 \xrightarrow[0°]{benzene} (C_6H_5)_3PCl_2 \xrightarrow[benzene]{1,2\text{-epoxycyclohexane}}$$

Submitted by James E. Oliver and Philip E. Sonnet[1]
Checked by Jerrold M. Liesch and George Büchi

1. Procedure

Caution! See benzene warning, p. 168. A hood should be employed for the chlorination.

A 1-l., three-necked flask is charged with 95 g. (0.36 mole) of triphenylphosphine (Note 1) and 500 ml. of anhydrous benzene. The flask is fitted with a gas inlet (Note 2), a mechanical stirrer, and a condenser with attached drying tube. The flask is cooled in an ice bath, stirring is begun, and chlorine is admitted through the gas inlet. Dichlorotriphenylphosphorane separates as a white solid or as a milky oil; the flow of chlorine is discontinued when the mixture develops a strong lemon-yellow color (Note 3). The gas inlet is quickly replaced by an addition funnel, and a solution of 10 g. of triphenylphosphine in 60 ml. of benzene is added dropwise fairly rapidly (Note 4). A solution of 24.5 g. (0.25 mole) of 1,2-epoxycyclohexane (Note 5) in 50 ml. of benzene is then added dropwise over *ca.* 20 minutes. The ice bath is replaced by a heating mantle, and the mixture, which consists of two liquid phases, is stirred and refluxed for 4 hours. It is then cooled, and excess dichlorotriphenylphosphorane is destroyed by the slow addition of 10 ml. of methanol (Note 6). The mixture is concentrated on a rotary evaporator at *ca.* 100 mm., and the residue, which may be a white solid or a viscous oil, is triturated with 300 ml. of petroleum ether (30–60°). The solid triphenylphosphine oxide that separates is collected by suction filtration. The cake is thoroughly broken up with a spatula, then washed with three 100-ml. portions of petroleum ether. The combined filtrates, from which a little more triphenylphosphine oxide precipitates, are refiltered, then washed with 250-ml. portions of aqueous 5% sodium bisulfite (Note 7) and with water. The organic phase is dried over magnesium sulfate, filtered, concentrated on a rotary evaporator at *ca.* 100 mm., and distilled through a 20-cm. Vigreux column. There is very little forerun, and *cis*-1,2-dichlorocyclohexane is collected at 105–110° (33 mm.), n^{25} D 1.4977. The yield is 27–28 g. (71–73%) (Note 8).

2. Notes

1. Triphenylphosphine was purchased from Aldrich Chemical Company, Inc. Use of a considerable excess of triphenylphosphine ensures complete reaction and obviates the need for rigorously dried glassware and reagents. Hydrochloric acid, generated by the reaction of dichlorotriphenylphosphorane and water, can react with the epoxide to produce a *trans*-chlorohydrin; *trans*-chlorohydrins, however, are converted to *cis*-dichlorides by dichlorotriphenylphosphorane under the conditions of the reaction.

2. A glass tube of 7-mm. diameter is recommended. If chlorine is introduced through a fritted-glass tube, the dichlorotriphenylphosphorane collects on the frit as a sticky gum.

3. A sharp endpoint is not observed. A simple test for complete chlorination is as follows: the flow of chlorine and the stirrer are stopped, and the mixture is allowed to settle. Chlorine is then admitted without stirring. If unreacted triphenylphosphine is present, a visible clouding (formation of dichlorotriphenylphosphorane) will occur at the gas–liquid interface.

4. Although a slight excess of chlorine does not appear to be deleterious, a substantial excess is avoided by adding the last portion of triphenylphosphine at this point.

5. Commercial 1,2-epoxycyclohexane, supplied by Columbia Organic Chemicals Company, Inc., was used.

6. The reaction mixture may be allowed to stand overnight before addition of the methanol.

7. The distilled *cis*-dichlorocyclohexane tends to become colored if the solution is not washed with a reducing agent.

8. The checkers, using a 10-cm. Vigreux column, found that it was necessary to take a wider boiling range fraction (105–115°,

33 mm.) to obtain similar yields. The product is virtually free of trans-1,2-dichlorocyclohexane (the isomeric 1,2-dichlorocyclohexanes are readily separated by gas chromatography on Carbowax 20M or on diethylene glycol succinate columns).

3. Discussion

This procedure is general for the conversion of epoxides to dichlorides with inversion of configuration at each of the two carbons and, in effect, provides a method for the cis-addition of chlorine to a double bond.[2] cis-1,2-Dichlorocyclohexane has also been prepared from 1,2-epoxycyclohexane and sulfuryl chloride,[3] but the stereospecificity of the reaction appears to be extremely sensitive to reaction conditions, and the yield is lower than that obtained by the method described here. Other methods give cis-1,2-dichlorocyclohexane contaminated with considerable amounts of the trans-isomer. This method has been used to convert cis- and trans-4,5-epoxyoctanes to meso- and dl-4,5-dichlorooctanes, respectively, and trans-7,8-epoxyoctadecane to threo-7,8-dichlorooctadecane. These conversions were carried out on smaller amounts of material, and the products were purified by column chromatography on silica gel. Yields were 51–63%.

Halogenations with dihalotriphenylphosphoranes have been reviewed briefly by Fieser and Fieser.[4] Dibromotriphenylphosphorane appears to have been studied somewhat more than the dichloro compound, but both reagents effectively convert alcohols to alkyl halides, carboxylic acids and esters to acid halides, etc. The reaction of 1,2-epoxycyclohexane with dibromotriphenylphosphorane under conditions similar to those described here gives a mixture of cis- and trans-1,2-dibromocyclohexanes. A reagent prepared from triphenylphosphine and carbon tetrachloride has been used for similar transformations.[5]

1. Agricultural Environmental Quality Institute, Agricultural Research Service, USDA, Beltsville, Maryland 20705. Mention of a proprietary product or company does not imply endorsement by the U.S. Department of Agriculture.
2. P. E. Sonnet and J. E. Oliver, J. Org. Chem., 41, 3279 (1976).
3. J. R. Campbell, J. K. N. Jones, and S. Wolfe, Can. J. Chem., 44, 2339 (1966).
4. L. F. Fieser and M. Fieser, "Reagents for Organic Synthesis," Vol. I, John Wiley & Sons, New York, N. Y., 1968, p. 1247.
5. J. G. Calzada and J. Hooz, Org. Syn., 54, 63 (1974).

Appendix
Chemical Abstracts Nomenclature (Collective Index Number;
Registry Numbers)

Cyclohexane, 1,2-dichloro-, *cis*- (8,9); (10498-35-8)

Phosphine, triphenyl- (8,9); (603-35-0)

Phosphorane, dichlorotriphenyl- (8,9); (2526-64-9)

1,2-Epoxycyclohexane: 7-Oxabicyclo[4.1.0]heptane (8,9); (286-20-4)

Phosphine oxide, triphenyl- (8,9); (791-28-6)

Cyclohexane, 1,2-dichloro-, *trans*- (8,9); (822-86-6)

Octane, 4,5-epoxy-, *cis*- (8); Oxirane, 2,3-dipropyl-, *cis*- (9); (1439-06-1)

Octane, 4,5-epoxy-, *trans*- (8); Oxirane, 2,3-dipropyl-, *trans*- (9); (1689-70-9)

meso-4,5-Dichlorooctane: Octane, 4,5-dichloro-, (R^*,S^*)- (8,9); (51149-23-6)

dl-4,5-Dichlorooctane: Octane, 4,5-dichloro-, (R^*,R^*)- (8,9); (51149-24-7)

trans-7,8-Epoxyoctadecane: Oxirane, 2-decyl-3-hexyl-, *trans*- (8,9); (59907-01-6)

threo-7,8-Dichlorooctadecane: Octadecane, 7,8-dichloro-, (R^*,R^*)- (8,9); (59840-26-5)

Phosphorane, dibromotriphenyl- (8,9); (1034-39-5)

Cyclohexane, 1,2-dibromo-, *cis*- (8,9); (19246-38-9)

Cyclohexane, 1,2-dibromo-, *trans*- (8,9); (7429-37-0)

2,3-DICYANOBUTADIENE AS A REACTIVE INTERMEDIATE
BY *in situ* GENERATION FROM 1,2-DICYANOCYCLOBUTENE:
2,3-DICYANO-1,4,4a,9a-TETRAHYDROFLUORENE

(2,3-Fluorenedicarbonitrile 1,4,4a,9a-tetrahydro-)

A.

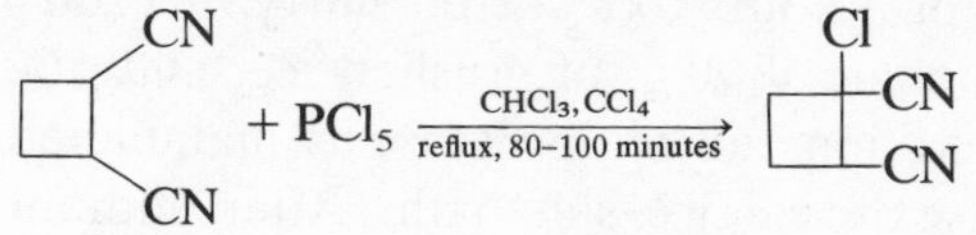

1

B.   **1**+(C$_2$H$_5$)$_3$N $\xrightarrow[\text{reflux}]{\text{benzene}}$

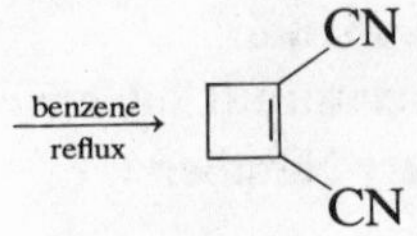

2

C.   **2** $\xrightarrow[\text{indene (hydroquinone)}]{150°, 4 \text{ hours}}$

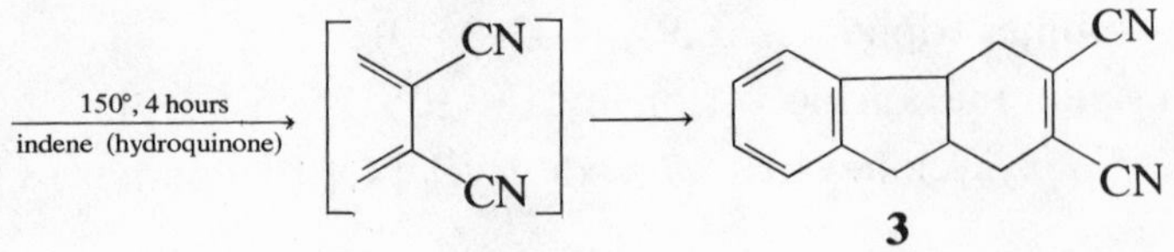

3

Submitted by D. Belluš,[1] H. Sauter, and C. D. Weis
Checked by A. J. Arduengo and William A. Sheppard

1. Procedure

Caution! See benzene warning, p. 168.

A. *1-Chloro*-1,2-*dicyanocyclobutane* (**1**). A 2-l., three-necked flask is equipped with a mechanical stirrer, a 500-ml. pressure-equalizing funnel, and an efficient reflux condenser provided with a gas outlet tube connected by plastic tubing to a conical funnel inverted over a 5-l. beaker containing aqueous sodium hydroxide for absorption of the evolved hydrogen chloride. The flask is charged with 562 g. (2.7 mole) of phosphorus pentachloride and 750 ml. of carbon tetrachloride. The rapidly stirred suspension is heated to reflux, and a solution of 159 g. (1.5 mole) of 1,2-dicyano-cyclobutane (Note 1) in 120 ml. of chloroform is added drop-wise over a period of 40 minutes (Note 2). After addition is complete, the solvents and the phosphorus trichloride formed during the reaction are removed by distillation at 100–150 mm. over a period of 40–60 minutes, with a bath temperature not exceeding 80° (Note 2). The residual liquid is cooled to room temperature and dissolved in 400 ml. of ethyl ether (Note 3). The etheral solution is placed in a 500-ml. dropping funnel and added over a period of 3 hours to a stirred slurry of 1700 g. (12 mole) of sodium bicarbonate, 800 g. of crushed ice, and 500 ml. of water. During the addition the temperature is maintained between −5° and 0° by an external ice–salt bath. After addition is complete, stirring is continued for 1 hour at the same temperature. The precipitated salts are removed by suction filtration through a sintered-glass funnel of medium porosity, and are thoroughly washed with 300 ml. of ethyl ether. The organic layer is separated and the aqueous filtrate extracted with three 200-ml. portions of ethyl ether. The combined ether extracts are dried over anhydrous magnesium sulfate, filtered, and concentrated on a rotary evaporator to yield 166–174 g. (79–83%) of a yellow oil consisting of an isomeric mixture of crude cyclobutane 1 (Note 4).

B. *1,2-Dicyanocyclobutene* (2). A 2-l., three-necked, round-bottomed flask is fitted with a 500-ml. pressure-equalizing dropping funnel, a mechanical stirrer, and a reflux condenser protected from moisture by a calcium chloride tube. The flask is charged with 131 g. (1.3 mole) of triethylamine (Note 5) and 400 ml. of benzene. The stirred solution is heated to gentle reflux, and a solution of 168.5 g. (1.2 mole) of crude cyclobutane 1 in 200 ml. of benzene is added dropwise over a period of 30 minutes. After addition is complete, the mixture is stirred under reflux for an additional 2 hours; the precipitated triethylamine hydrochloride is then filtered from the cold solution and washed with 150 ml. of benzene. The combined filtrates are washed twice with 200-ml. portions of water and evaporated using a water aspirator at a bath temperature of 35°. The residue is distilled to yield 105–110 g. (83–87%) of crude cyclobutene 2, b.p. 55–60° (0.06 mm.) (Note 6).

C. *2,3-Dicyano-1,4,4a,9a-tetrahydrofluorene* (3). A 100-ml., round-bottomed flask, equipped with a reflux condenser under nitrogen pressure, is charged with 10.4 g. (0.10 mole) of crude cyclobutene 2, 23.3 g. (0.20 mole) of indene (Note 7), and 0.3 g. of hydroquinone. The reaction mixture is stirred and heated at 150° for 4 hours under nitrogen. The reflux condenser is then replaced by a still head, and 6.3 g. (0.053 mole) of indene is distilled from the flask at a bath temperature of about 95° and a pressure of 11 mm. (Note 8). The dark-colored reaction mixture is transferred to a 500-ml., round-bottomed flask, diluted with 200 ml. of benzene followed by 1 g. of decolorizing carbon, and the resulting mixture is refluxed for 2 hours. After the mixture is cooled to room temperature, the carbon is removed by filtration, and the benzene is distilled. The residual oily residue solidifies on standing and is recrystallized from 45 ml. of ethanol to yield 15.4–15.9 g. (70–72%) of crystalline fluorene 3, m.p. 98.5–100° (Note 9).

2. Notes

1. 1,2-Dicyanocyclobutane (*cis*- and *trans*-isomer mixture) was purchased from Aldrich Chemical Company, Inc., and used without further purification.

2. The addition and distillation must be accomplished within the specified period of time; otherwise the amount of dichlorinated 1,2-dicyanocyclobutane increases considerably. The submitters found that an 80% molar excess of phosphorus pentachloride is optimum. A molar excess less than specified (under given experimental conditions) gives considerable unreacted starting material. Under forcing experimental conditions, such as longer reaction times and/or higher temperatures, the starting material can be completely consumed, even with less than 80% molar excess of PCl_5, but a considerable amount of dichlorinated products is formed.

3. The checkers found that, because of the time required for completion of this step, a convenient modification is to cool the ethyl ether solution to −78° in dry ice and to store overnight at −78°. A solid complex of phosphorus pentachloride and cyclobutane **1**, m.p. 88–91°, precipitates in the cold ethyl ether solution. This complex may not redissolve on warming to room temperature, but the suspension in ethyl ether can be used to proceed with the second half of Step A.

4. The checkers obtained the cyclobutane **1** as a colorless crystalline solid, m.p. 45–47° (a mixture of major and minor isomers), that is relatively free of 1,2-dichloro-1,2-dicyanocyclobutane. The product had the following spectral properties: proton magnetic resonance (chloroform-*d*) δ (multiplicity): 2.35–3.15 (multiplet), 3.40–4.20 (multiplet); ^{13}C magnetic resonance (chloroform-*d*) δ: major isomer 21.58 (triplet), 38.41 (doublet), 37.24 (triplet), minor isomer 22.48 (triplet), 36.72 (doublet), 36.91 (triplet).

5. Reagent-grade triethylamine was dried over sodium hydroxide and distilled before use (b.p. 88–89°).

6. The checkers obtained yields of 115 g. (92%) on a 1.2 mole scale and 94 g. (90%) on a 1.0 mole scale.

The product was analyzed by gas chromatography on a 1.23 m.× 0.65 cm. stainless-steel column of SE-52 on Varoport 30, which was heated to 150° and swept with helium at 60 ml. per minute.

Retention times for the various components (minutes) are: cy-clobutene **2**, 2.6; *trans*-1,2-dicyanocyclobutane, 3.6; two isomeric 1,2-dichloro-1,2-dicyanocyclobutanes, 5.1 and 5.9, respectively; *cis*-1,2-dicyanocyclobutane, 9.8.

For most synthetic purposes, such as [4+2]- and [2+2]-cycloadditions,[2-7] ring-opening reactions,[2,8] and hydrolytic reactions,[2] this crude cyclobutene **2**, which contains approximately 1–3.5% of isomeric mixtures of 1,2-dichloro-1,2-dicyano-cyclobutanes, can be used satisfactorily without further purification. Pure cyclobutene **2** can be prepared by treatment of the crude product with Raney cobalt, thereby removing residual quantities of isomeric 1,2-dichloro-1,2-dicyanocyclobutanes. In a typical experiment the crude product is placed in a 250-ml., round-bottomed flask and stirred with 10 g. of Raney cobalt for 4 hours at 70° under nitrogen. Distillation directly from the reaction vessel without filtering off the metal slurry yields 94–98 g. (60–63%) of cyclobutene **2** as a colorless liquid, n^{20}D 1.4926, d^4_{20} 1.033. The Raney cobalt used by the submitters was obtained from Fluka A G, Buchs, Switzerland, as a suspension in water, and was washed with tetrahydrofuran before use. Raney nickel and nickel tetracarbonyl, respectively, are also good dechlorinating reagents. The use of Raney cobalt, however, diminishes the danger of self-ignition during the preparation procedure. The spectral properties are as follows: infrared (neat) cm^{-1}: 3002, 2957, 2230, 1612, 1422, 1251, 1169, 1003, and 623; ultraviolet (methanol) nm. max. (log ε): 235 (4.06), 247 (shoulder, 3.90); proton magnetic resonance (chloroform-d) δ (multiplicity): 2.91 (singlet).

7. The indene used by the submitters was "practical grade," purchased from Fluka A G, Buchs, Switzerland. The indene used by the checkers was purchased from Aldrich Chemical Company, Inc. Both were distilled (b.p. 60–65°, 11 mm.) before use.

8. Recovered indene may be used for the next batch without further purification.

9. Fluorene **3** has the following spectral properties: infrared (KBr) cm^{-1}: 2222, 1608, 1479, 1440, 773, and 742; ultraviolet (ethanol) nm. max. (log ε): 216 (4.10), 2.34 (3.99), 261 (3.16), 267 (3.11), and 274 (3.07); mass spectrum m/e: 220 (m^+) and 116 (base peak); proton magnetic resonance (100 MHz., chloroform-d): two complex multiplets for the aromatic and aliphatic protons

(360 MHz., chloroform-d) δ (multiplicity, coupling constant J in Hz., number of protons, assignment): 2.10 [doublet $J = 18$, doublet

$J = 7$, and triplet $J = 3$, one H of $CH_2(1)$ or one H of $CH_2(4)$], 2.55–2.70 [multiplet, one H of $CH_2(1)$, one H of $CH_2(4)$, one H of $CH_2(9)$ and $CH(9a)$], 2.88 [doublet $J = 18$, doublet $J = 7$, doublet $J = 3$, and doublet $J = 1.5$, one H of $CH_2(1)$ or one H of $CH_2(4)$], 3.11 [doublet $J = 15$ and doublet $J = 6$, one H of $CH_2(9)$], 3.42 [doublet $J = 5$ and triplet $J = 7$, $CH(4a)$], 7.20 (multiplet, 4, aromatic H); ^{13}C magnetic resonance spectrum (chloroform-d) δ (assignment): 143.8 [$C(8a)$], 141.3 [$C(4b)$], 127.6, 127.2 and 125.6 [$C(6)$, $C(7)$, and $C(8)$, not assigned individually], 126.6 and 125.9 [$C(2)$ and $C(3)$, not assigned], 123.2 [$C(5)$], 115.9 [two nitrile carbons], 40.2 [$C(4a)$], 38.4 [$C(9)$], 35.4 [$C(9a)$], 30.7 and 29.7 [$C(1)$ and $C(4)$, not assigned].

3. Discussion

Three syntheses of 1,2-dicyanocyclobutene (**2**) have been previously described. The first, involving dehydration of cyclobutene-1,2-dicarboxamide does not specify the yield.[9] The second procedure involves a concomitant chlorination and catalytic dehydrochlorination of 1,2-dicyanocyclobutane in the gas phase, yielding 1,2-dicyanocyclobutene (**2**) in a mixture of several other products.[10] The third method consists of dechlorination of 1,2-dichloro-1,2-dicyanocyclobutane using metals, such as zinc copper couple,[11] Raney nickel,[11] and, especially, Raney cobalt.[2] In comparison with this last-mentioned synthesis, the overall yield of the present procedure is 5–10% higher. Furthermore, the reaction is performed in less time and utilizes considerably cheaper reagents.

Pure crystalline 2,3-dicyanobutadiene has been prepared in high yield by gas-phase thermolysis of cyclobutene (**2**).[2,8] Analogous thermolysis of derivatives of cyclobutene-1,2-dicarboxylic acid appears to represent general procedures for the synthesis of derivatives of butadiene-2,3-dicarboxylic acid of high purity.[2,12] These

TABLE I
[4 + 2]-CYCLOADDITION REACTIONS OF 2,3-DICYANOBUTADIENE FORMED *in situ*
FROM 1,2-DICYANOCYCLOBUTENE[2]

Olefin	Product	Yield (%)[a]	Temperature (°C)	Time (hours)
Norbornadiene	4,5-Dicyanotricyclo-[6.2.1.0^{2,7}]undeca-4,9-diene	79[b]	150	12
Acenaphthylene	8,9-Dicyano-6b,7,10,10a-tetrahydrofluoranthene	77	138	48
Cyclopentene	3,4-Dicyanobicyclo[4.3.0]-non-3-ene	65	135	16
Ethylene	1,2-Dicyanocyclohexene	58	135	16
(E)-Stilbene	1,2-Dicyano-4,5-diphenyl-cyclohexene	32	138	24
Butyl vinyl ether	1,2-Dicyano-4-butoxy-cyclohexene	28	155	16
(E)-1,2-Dichloroethylene	1,2-Dicyano-4,5-*trans*-dichlorocyclohexene	8	135	16
2-Vinylpyridine	1,2-Dicyano-4-(2′-pyridyl)-cyclohexene	5	138	48

[a] Yields of analytically pure products are given.
[b] A 70:24 mixture of *exo* and *endo* isomers. Some 2:1 cycloadduct was also isolated (2.4% yield).

butadienes take part in [4 + 2]-cycloaddition reactions either as reactive dienes[2,13,14] or as reactive dienophiles.[2,14] In the pure state, however, they tend to polymerize, and even crystalline 2,3-dicyanobutadiene slowly polymerizes to yield a highly cross-linked polymer without losing its original crystal form. A [2 + 4]-dimer of a 2,3-dicyanobutadiene is also formed by heating the dicyanocyclobutene in solution with a polymerization inhibitor.[2] Monomeric derivatives of butadiene-2,3-dicarboxylic acid cannot be prepared in solution due to rapid dimerization.[8,14]

The present procedure of *in situ* generation and trapping of 2,3-dicyanobutadiene in the presence of olefins overcomes these problems and affords the [4 + 2]-cycloadducts in good yields, particularly in the case of olefins possessing a strained double bond.[2] Substituted 1,2-dicyanocyclohexenes prepared by the *in situ* [4 + 2]-cycloadditions can be dehydrogenated to new aromatic *ortho*-dinitriles. For example, 2,3-dicyanofluorene is prepared in

56% yield by heating 2,3-dicyano-1,4,4a,9a-tetrahydrofluorene (3) at 200° in dimethylmaleate in the presence of 5% palladium on charcoal. Other aromatic *ortho*-dinitriles have also been prepared by this method.[2] Because 2,3-dicyanobutadiene is an electron-deficient diene, it does not react with electron-deficient olefins, such as maleic anhydride and fumaronitrile,[2,8] with this procedure. However, by generating the dicyanobutadiene in refluxing chlorobenzene in the presence of maleic anhydride with 2,5-di-*tert*-butylbenzoquinone inhibitor, the [2 + 4]-cyclic dicarbonitrile adduct, m.p. 201–202.5°, was formed in a yield of 38%.[15]

1. Central Research Laboratories, Ciba-Geigy A G, CH-4002 Basel, Switzerland.
2. D. Belluš, K. von Bredow, H. Sauter, and C. D. Weis, *Helv. Chim. Acta*, **56**, 3004 (1973).
3. D. Belluš and G. Rist, *Helv. Chim. Acta*, **57**, 194 (1974).
4. D. Belluš, H.-C. Mez, and G. Rihs, *J. Chem. Soc., Perkin Trans. II*, 884 (1974).
5. D. Belluš, H.-C. Mez, G. Rihs, and H. Sauter, *J. Amer. Chem. Soc.*, **96**, 5007 (1974).
6. R. Wehrli, H. Schmid, D. Belluš, and H.-J. Hansen, *Helv. Chim. Acta*, **60**, 1325 (1977).
7. H.-D. Martin, M. Hekman, G. Rist, H. Sauter, and D. Belluš, *Angew. Chem.*, **89**, 420 (1977).
8. D. Belluš and C. D. Weis, *Tetrahedron Lett.*, 999 (1973).
9. H. Prinzbach and H.-D. Martin, *Chimia*, **23**, 37 (1969).
10. J. L. Greene and M. Godfrey, U.S. Patent 3,336,354 (1967) [*C.A.*, **68**, 21598v (1968)].
11. J. L. Greene, N. W. Standish, and N. R. Gray, U.S. Patent 3,275,676 (1966) [*C.A.*, **66**, 10637y (1967)].
12. P. Dowd and K. Kang, *Synth. Commun.*, **4**, 151 (1974).
13. E. Vogel, *Justus Liebigs Ann. Chem.*, **615**, 14 (1958).
14. U.-I. Záhorszky and H. Musso, *Justus Liebigs Ann. Chem.*, 1777 (1973).
15. W. A. Sheppard, unpublished results.

Appendix
Chemical Abstracts Nomenclature (Collective Index Number; Registry Numbers)

1,2-Dicyanocyclobutane: 1,2-Cyclobutanedicarbonitrile (8,9); (3396-17-6)

1-Chloro-1,2-dicyanocyclobutane: 1,2-Cyclobutanedicarbonitrile, 1-chloro- (8,9); (3716-98-1)

1,2-Dichloro-1,2-dicyanocyclobutane: 1,2-Cyclobutanedicarbonitrile, 1,2-dichloro- (8,9); (3496-67-1)

1,2-Dicyanocyclobutene: 1-Cyclobutene-1,2-dicarbonitrile (8,9); (3716-97-0)

2,3-Dicyanobutadiene: Succinonitrile, dimethylene- (8); Butane-dinitrile, bis(methylene)- (9); (19652-57-4)

2,3-Dicyanofluorene: 2,3-Fuorenedicarbonitrile (8); 9H-Fluo-rene-2,3-dicarbonitrile (9); (52477-74-4)

Cyclobutene-1,2-dicarboxylic acid: 1-Cyclobutene-1,2-dicarbox-ylic acid (8,9); (16508-05-7)

Cyclobutene-1,2-dicarboxamide: 1-Cyclobutene-1,2-dicarbox-amide (8,9); (23335-15-1)

Butadiene-2,3-dicarboxylic acid: Succinic acid, dimethylene- (8); Butanedioic acid, bis(methylene)- (9); (488-20-0)

FLUORINATIONS WITH PYRIDINIUM POLYHYDROGEN FLUORIDE REAGENT: 1-FLUOROADAMANTANE

(Tricyclo[3.3.1.1^{3,7}]decane, 1-fluoro-)

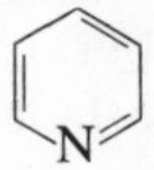

Submitted by GEORGE A. OLAH and MICHAEL WATKINS[1]
Checked by PAUL G. WILLIARD and G. BÜCHI

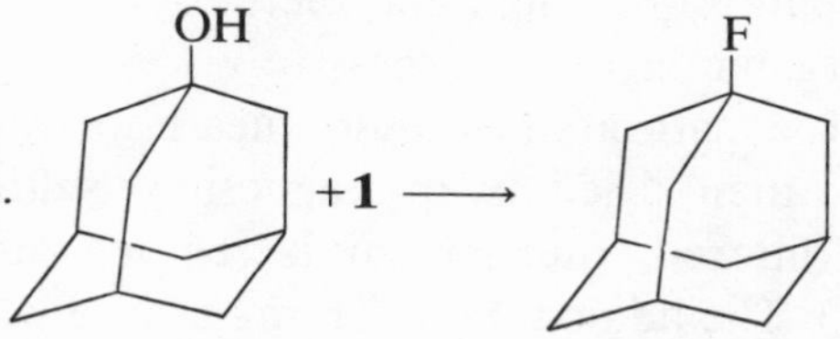

1. Procedure

Caution! Proper precautions must be used when handling anhy-drous hydrogen fluoride and pyridinium polyhydrogen fluoride. Hy-drogen fluoride is extremely corrosive to human tissue, contact resul-ting in painful, slow-healing burns. Laboratory work with HF should be conducted only in an efficient hood, with the operator wearing a full-face shield and protective clothing (Note 1).

A. *Pyridinium polyhydrogen fluoride* (**1**). A previously tared 250-ml. polyolefin bottle is equipped with a polyolefin gas inlet and drying tube inserted through holes in the cap and sealed with Teflon tape. The bottle is charged with 37.5 g. (0.475 mole) of pyridine (Note 2) and is cooled in an acetone–dry ice bath. After the pyridine has solidified, 87.5 g. (4.37 moles) of anhydrous hydrogen fluoride (Note 3) is condensed from a cylinder into the bottle through the polyolefin inlet tube. The amount of hydrogen fluoride is determined by weighing the bottle. After the hydrogen fluoride has cooled, the bottle is cautiously swirled with cooling until the solid dissolves (Note 4). The solution can now be safely allowed to warm to room temperature.

B. *1-Fluoroadamantane* (**2**). A 250-ml. polyolefin bottle is equipped with a Teflon-coated magnetic stirring bar and a polyolefin drying tube inserted through a hole in the cap and sealed with Teflon tape. The bottle is charged with 5.0 g. (0.033 mole) of 1-adamantanol (Note 5) and 50 ml. of pyridinium polyhydrogen fluoride (**1**). The solution is allowed to stir for 3 hours at ambient temperature. After this period 150 ml. of petroleum ether is added, and the stirring is continued for another 15 minutes. The resulting two-phase solution is transferred to a 250-ml. polyolefin separatory funnel, and the bottom layer is discarded (Note 6). The organic layer is successively washed with 50 ml. of water, 50 ml. of a saturated sodium bicarbonate solution, and 50 ml. of water, then dried over magnesium sulfate. After the organic layer is filtered, the solvent is removed under reduced pressure (Note 7). The adamantane **2** remains as a white powder in a yield of 4.5–4.6 g. (88–90%), m.p. 225–227° (sublimes in sealed capillary) (Note 8). Further purification, if needed, is by vacuum sublimation or by recrystallization from a mixture of methanol and carbon tetrachloride.

2. Notes

1. The recommended procedure for an HF burn is to flood with water, pack with ice, and get medical attention as quickly as possible. Local medical personnel should be alerted and prepared when work with HF is planned. Directions for proper medical treatment are given in reference 2.

2. The submitters used A.C.S. certified reagent-grade pyridine from Fisher Scientific Company which was distilled from potassium hydroxide prior to use.

3. The submitters obtained anhydrous HF from Harshaw Chemical Company, and the checkers purchased this reagent from Matheson Gas Products.

4. The dissolution is an extremely exothermic process that can be violent if the bath temperature is not carefully controlled at −78°. A preferred procedure developed by Dr. A. E. Feiring, Central Research and Development Department, Du Pont Experimental Station, involves keeping the pyridine as cold as possible without freezing (ca. −40°), then slowly condensing the HF into the vessel so that the entire mixture remains liquid during the preparation. Stirring is also helpful.

5. This was obtained from Aldrich Chemical Company, Inc., and used without further purification.

6. The inorganic layer can be safely disposed by slow addition to large amounts of ice-cold water.

7. Since 1-fluoroadamantane sublimes easily, the water bath should be controlled to about 32°, and the vacuum evaporation of the solvent limited to as short a time as possible.

8. Gas-chromatographic and infrared analysis indicate no detectable amount of starting alcohol. Proton magnetic resonance of adamantane 2 (chloroform-d) yields a series of multiplets centered at δ 1.62, 1.86, 2.18.

3. Discussion

1-Fluoroadamantane (2) can also be prepared by halogen exchange from the bromide with silver fluoride[3] or with zinc fluoride.[4] The procedure described is a more convenient and economical method than the halogen exchanges. It has found successful application in the preparation of a wide variety of secondary and tertiary fluorides from their corresponding alcohols, with yields generally falling in the range of 70–90%[5] (Table I).

The HF–pyridine reagent is an effective complement to dimethylaminosulfur trifluoride (DAST) reagent[6] in the preparation of alkyl fluorides from alcohols. DAST is also useful for the conversion of carbonyl groups to difluoromethylene functions. The

TABLE I

PREPARATION OF TERTIARY- AND SECONDARY-ALKYL FLUORIDES FROM
ALCOHOLS WITH HYDROGEN FLUORIDE–PYRIDINE REAGENT

Alcohol	Temp- erature	Reaction Time (hours)	Alkyl Fluoride	b.p. (m.p.)	Yield (%)
Isopropyl	50	3.0	Isopropyl	−9 to −7	30
sec-Butyl	20	3.0	sec-Butyl	25–26	70
tert-Butyl	0	1.0	tert-Butyl	12	50
3-Ethyl-3-pentyl	0	0.5	3-Ethyl-3-pentyl	30–33 (60 mm.)	95
3-Methyl-3-heptyl	−70	0.5	3-Methyl-3-heptyl	35 (40 mm.)	85
3-Methyl-4-heptyl	0	2.0	3-Methyl-4-heptyl		35
Cyclohexyl	20	2.0	Cyclohexyl fluoride	100–102	99
2-Norbornyl	20	1.0	2-Norbornyl	(56–59)	95
2-Adamantyl	20	0.5	2-Adamantyl	(254–255)	98
α-Phenylethyl	20	0.5	α-Phenylethyl	46 (15 mm.)	65

HF–pyridine reagent, however, can also be used for the hydrofluorination of alkenes,[7] alkynes,[7] cyclopropanes,[7] and diazo compounds,[8] the halofluorination of alkenes,[9] the preparation of fluoroformates from carbamates,[10] the preparation of α-fluorocarboxylic acids from α-amino acids,[11] and as a deprotecting reagent in peptide chemistry.[12] Examples of the hydrofluorination of alkenes with HF–pyridine are given in Table II.

TABLE II

HYDROFLUORINATION OF ALKENES WITH HYDROGEN FLUORIDE–PYRIDINE
REAGENT

Alkene	Reaction Temp- erature	Product	b.p. (m.p.)	Yield (%)
Propene	20	Isopropyl fluoride	−11 to −9	35
Cyclopropane	20	Propyl fluoride	−3 to −1	75
2-Butene	0	sec-Butyl fluoride	24–25	40
2-Methylpropene	0	tert-Butyl fluoride	11–13	60
Cyclopentene	0	Cyclopentyl fluoride	51–52 (200 mm.)	65
Cyclohexene	0	Cyclohexyl fluoride	102–104	80
Cycloheptene	0	Cycloheptyl fluoride	70–71 (200 mm.)	90
Norbornene	0	2-Norbornyl fluoride	(56–59)	65
1-Hexyne	0	2,2-Difluorohexane	85–87	70
3-Hexyne	0	3,3-Difluorohexane	84–86	75

1. Institute of Hydrocarbon Chemistry, Department of Chemistry, University of Southern California, Los Angeles, California 90007.
2. G. A. Olah and S. J. Kuhn, *Org. Syn.*, Coll. Vol. **5**, 66 (1973); C. M. Sharts and W. A. Sheppard, *Org. React.*, **21**, 192, 220–223 (1974).
3. R. C. Fort and P. v. R. Schleyer, *J. Org. Chem.*, **30**, 789 (1965).
4. K. S. Bhandari and R. E. Pincock, *Synthesis*, 655 (1977).
5. G. A. Olah, M. Nojima, and I. Kerekes, *Synthesis*, 786 (1973).
6. W. J. Middleton and E. M. Bingham, *Org. Syn.*, **57**, 50, 72 (1977).
7. G. A. Olah, M. Nojima, and I. Kerekes, *Synthesis*, 779 (1973).
8. G. A. Olah and J. Welch, *Synthesis*, 896 (1974).
9. G. A. Olah, M. Nojima, and I. Kerekes, *Synthesis*, 780 (1973).
10. G. A. Olah and J. Welch, *Synthesis*, 654 (1974).
11. G. A. Olah and J. Welch, *Synthesis*, 652 (1974).
12. S. Matsuura, C. H. Niu, and J. S. Cohen, *Chem. Commun.*, 451 (1976).

Appendix
Chemical Abstracts Nomenclature (Collective Index Number; Registry Numbers)

Adamantane, 1-fluoro- (8); Tricyclo[3.3.1.1^{3,7}]decane, 1-fluoro- (9); (768-92-3)

Hydrogen fluoride: Hydrofluoric acid (8,9); (7664-39-3)

Pyridine (8,9); (110-86-1)

1-Adamantanol (8); Tricyclo[3.3.1.1^{3,7}]decan-1-ol (9); (768-95-6)

γ-KETOESTERS FROM ALDEHYDES VIA DIETHYL ACYLSUCCINATES: 4-OXOHEXANOIC ACID ETHYL ESTER

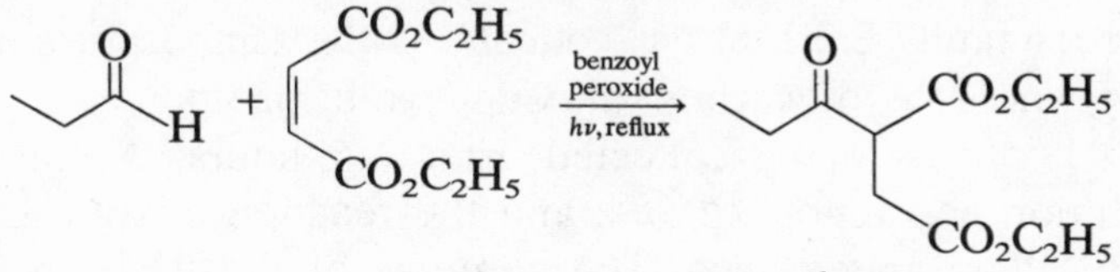

Submitted by PIUS A. WEHRLI and VERA CHU[1]
Checked by WILLIAM B. FARNHAM and WILLIAM A. SHEPPARD

1. Procedure

A. *Diethyl propionylsuccinate* (**1**). A solution of 412 g. (2.4 mole) of diethyl maleate (Note 1), 278 g. (4.8 mole) of freshly distilled propionaldehyde (Note 2), and 1.2 g. (0.0048 mole) of benzoyl peroxide in a normal 2-l. Pyrex flask is heated under reflux while undergoing irradiation with an ultraviolet lamp (Note 3). The initial reflux temperature is 60°. After 2 hours another 1.2 g. (0.0048 mole) of benzoyl peroxide is added. Strong reflux and irradiation are maintained throughout the entire reaction period. After 18 hours' total time, the internal pot temperature reaches 68°. At this point the last 1.2 g. (0.0048 mole) of benzoyl peroxide is added, and the reaction is continued for a total of 30 hours, at which time the pot temperature reaches 74.5°. The reflux condenser is then replaced by a distillation head. The excess propionaldehyde (119 g.) is distilled under atmospheric pressure, b.p. 48–49°. Succinate **1** is distilled under reduced pressure. The main fraction, b.p. 145–151.5° (15–16 mm), provides 417–449 g. (75–81%) of product having sufficient purity for use in the next step (Note 4).

B. *4-Oxohexanoic acid ethyl ester* (**2**). A 1-l., three-necked, round-bottomed flask is equipped with a mechanical stirrer, thermometer, and Claisen condenser connected to a gas-measuring device (Note 5). The flask is charged with 276 g. (1.2 mole) of succinate **1** and 74.1 g. (1.2 mole) of boric acid (Note 6). The initially heterogeneous mixture is stirred and immersed in a 150° oil bath. Within 1 hour 36 g. of distillate (mainly ethanol) and approximately 2.3 l. of gas collect. As the temperature is raised to 170°, the rate of carbon dioxide evolution increases, a total of 24.9 l. of gas being collected after 1.5 hours. At this time gas evolution has almost ceased, and the reaction mixture has a clear, light yellow appearance. The contents of the flask are cooled to room temperature, poured onto 1.5 l. of ice, and extracted with three 500-ml. portions of toluene. The combined organic layers are dried over anhydrous magnesium sulfate, and the solvent is removed under reduced pressure. The product is distilled through a 10-cm. Vigreux column to yield 156–162 g. (82–85%) of oxohexanate ester **2**, b.p. 109–112° (18 mm.). Gas chromatographic analysis indicates the material to be 99.2% pure (Note 7).

2. Notes

1. Diethyl maleate, practical grade, available from Eastman Organic Chemicals, was used without further purification.

2. Propionaldehyde was obtained from Aldrich Chemical Company, Inc. The aldehyde must be distilled before use.

3. The checkers used a 275-W. General Electric sunlamp. The submitters used a 140-W. Hanovia Ultraviolet Quartz lamp of a type no longer available.

4. The succinate **1** has the following proton magnetic resonance spectra (chloroform-d) δ (multiplicity, number of protons, assignment): 1.0–1.45 (multiplet, 9, CH_3), 2.6–3.05 (multiplet, 4, CH_2—$\overset{\overset{\displaystyle O}{\|}}{C}$), 3.9–4.4 (multiplet, 5, OCH_2 and tertiary CH).

5. For the gas-measuring device, the submitters used an inverted, calibrated 10-l. bottle, filled with saturated sodium chloride solution, resting in an enamel bucket big enough to hold the volume to be displaced. The checkers used a gas meter. However, the rate of gas evolution can be estimated by using a simple gas bubbler.

6. Reagent-grade boric acid, available from Aldrich Chemical Company, Inc., was used.

7. Gas chromatographic analysis was performed on a Hewlett-Packard Model 5720 with dual flame detector; column 1.85 m.× 0.313 cm. outer diameter, stainless steel; 10% UCW-98 on Diatoport 5, programmed at 30° per minute from 50–250°. The purity was calculated on an area comparison. The oxohexanate ester **2** has the following proton magnetic resonance spectrum (chloroform-d) δ (multiplicity, number of protons, assignment, coupling constant J in Hz.): 1.05 (triplet, 3, CH_3, $J=7$), 1.2 (triplet, 3, CH_3, $J=7$), 2.4–2.9 (multiplet, 6, CH_2), 4.2 (quartet, 2, OCH_2, $J=7$).

3. Discussion

γ-Ketoesters, notably 5-substituted ethyl levulinates, have been prepared via radical addition of aldehydes to diethyl maleate to give acylated diethyl succinates.[2] These intermediates in turn had to be saponified,[2] decarboxylated,[2] and reesterified to give the corresponding 4-oxocarboxylic acid esters. A more direct alternative method[3] utilizes the free radical addition of butyraldehyde to methyl acrylate, but the reported yield is low (11%).

The present method[4] is simple, versatile, and efficient in contrast to earlier methods, which were multistep or preparatively unsatisfactory. Various 5-substituted 4-oxocarboxylic acid esters can be prepared by this procedure.[4]

γ-Ketoesters in general and levulinic acid or esters in particular have extensive utility.[5] For example, they can serve as central intermediates for γ-butyrolactones,[6] 1,4-diols,[7] thiophenes,[8] pyrrolidones,[9] and 2-alkyl-1,3-cyclopentanediones.[10]

1. Chemical Research Department, Hoffmann–La Roche Inc., Nutley, New Jersey 07110.
2. T. M. Patrick, Jr., J. Org. Chem., 17, 1009 (1952).
3. E. C. Ladd, U.S. Patent 2,577,133 (1951) [C.A., 46, 6147h (1952)].
4. P. A. Wehrli and V. Chu, J. Org. Chem., 38, 3436 (1973).
5. R. H. Leonard, Ind. Eng. Chem., 48, 1330 (1956).
6. H. A. Schuette and P. Sah, J. Amer. Chem. Soc., 48, 3163 (1926).
7. A. Müller and H. Wachs, Monatsh. Chem., 53, 420 (1929).
8. N. R. Chakrabarty and S. K. Mitra, J. Chem. Soc., 1385 (1940).
9. R. L. Frank, W. R. Schmitz, and B. Zeidman, Org. Syn., Coll. Vol. 3, 328 (1955).
10. U. Hengartner and V. Chu, Org. Syn., 58, 83 (1978).

Appendix
Chemical Abstracts Nomenclature; (Collective Index Number; Registry Numbers)

Diethyl propionylsuccinate: Succinic Acid, propionyl-, diethyl ester (8); Butanedioic Acid, 1-oxopropyl-, diethyl ester (9); (41117-76-4)

Diethyl Maleate: Maleic Acid, diethyl ester (8); 2-Butenedioic Acid, diethyl ester (9); (141-05-9)

Propionaldehyde (8); Propanal (9); (123-38-6)

Benzoyl peroxide (8); Peroxide, dibenzoyl- (9); (94-36-0)

Hexanoic Acid, 4-oxo-, ethyl ester (8,9); (3249-33-0)

γ-KETOESTERS TO PREPARE CYCLIC DIKETONES: 2-METHYL-1,3-CYCLOPENTANEDIONE

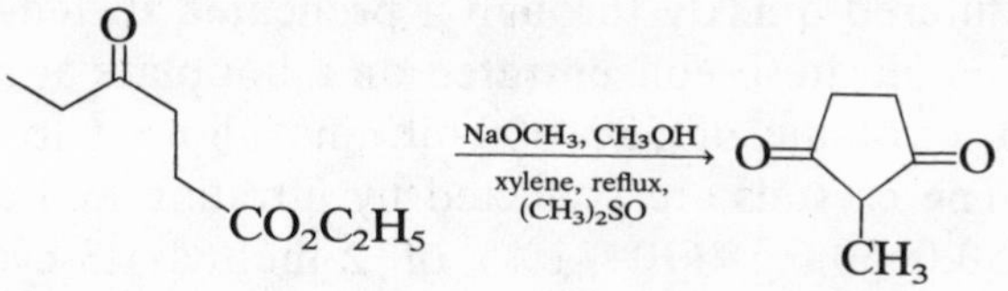

Submitted by U. Hengartner and Vera Chu[1]
Checked by William B. Farnham and William A. Sheppard

1. Procedure

A 3-l., three-necked, round-bottomed flask is fitted with a dropping funnel (Note 1), a mechanical stirrer, and a distillation head with a thermometer and efficient Liebig condenser. The flask is charged with 1.4 l. of xylene (Note 2). The xylene is stirred and heated to boiling with a heating mantle while 179 g. of a solution containing 43 g. (0.80 mole) of sodium methoxide in methyl alcohol (Note 3) is added over 20 minutes. During this period 450 ml. of solvent is distilled.

After addition is complete, 300 ml. of xylene is added and the distillation continued until the vapor temperature rises again to 138°, during which time an additional 250 ml. of distillate collects. This leaves a white suspension, to which 18 ml. of dimethylsulfoxide is added. Then a solution of 100 g. (0.633 mole) of 4-oxohexanoic acid ethyl ester (Note 4) in 200 ml. of xylene is added (Note 1) to the vigorously stirred sodium methoxide suspension over 25 minutes, while 900 ml. of distillate is collected continuously while the vapor temperature is maintained at 134–137°. The orange-colored mixture is stirred and heated for an additional 5 minutes and then cooled to room temperature. Addition of 165 ml. of water with vigorous stirring over a 5-minute period (Note 5) gives two clear phases, which are cooled in an ice bath and acidified by adding 82 ml. (0.98 mole) of 12N hydrochloric acid with vigorous stirring. After the mixture is stirred at 0° for another 1.5 hours, the crystalline product is collected by suction filtration and carefully washed successively with 100-ml. and 50-ml. portions of ice-cooled ethyl ether (Note 6).

The crude product is dissolved in 1 l. of boiling water, and the solution is filtered quickly through a preheated fritted-disk funnel (Note 7). The filtrate is concentrated on a hot plate at atmospheric pressure to a volume of 550–600 ml. and allowed to stand at 0° overnight. The crystals are collected by filtration and dried at 85° to afford 50.0–50.6 g. (70–71%) of 2-methyl-1,3-cyclopentane-dione, m.p. 210–211° (Note 8).

2. Notes

1. A 500-ml. dropping funnel, without pressure-equalizing arm but provided with a calcium sulfate drying tube, is used.

2. Reagent-grade xylene, b.p. 138–141°, from Fisher Scientific Company, was used.

3. The sodium methoxide solution is prepared as follows: 203 g. of methyl alcohol, available from Fisher Scientific Company, is placed in a 500-ml., two-necked flask under an inert atmosphere. The flask is equipped with a magnetic stirring bar and a reflux condenser provided with a calcium sulfate drying tube. Freshly cut, clean sodium (23 g., 1 mole) is added in small pieces at such a rate that reflux is maintained. The mixture is stirred until all the sodium has reacted.

4. 4-Oxohexanoic acid ethyl ester was prepared by the method of P. A. Wehrli and V. Chu, *Org. Syn.*, **58**, 79 (1978).

5. The temperature of the mixture is kept at 25–35°.

6. The ether washings contain 11 g. of a brown viscous oil containing various condensation products.

7. A small amount of insoluble tarry material is removed by this filtration. Preheating the funnel is necessary, since the product crystallizes easily on cooling. Dark-colored impurities in crude 2-methyl-1,3-cyclopentanedione can be removed by recrystallization from methanol.

8. 2-Methyl-1,3-cyclopentanedione, literature[6] m.p. 210–212°, exists in the enol form in solution. It has the following spectral properties: ultraviolet (0.1N HCl) nm. max. (ε): 252 (19,000); proton magnetic resonance (dimethyl sulfoxide-d_6) δ(multiplicity, number of protons, assignment): 1.5 l. (singlet, 3, CH_3), 2.39 (singlet, 4, CH_2).

3. Discussion

2-Methyl-1,3-cyclopentanedione is a key intermediate for the total synthesis of steroids.[2] A number of methods have been described for its preparation, among them the condensation of succinic acid with propionyl chloride,[3] and that of succinic anhydride with 2-buten-2-ol acetate,[4] both in the presence of aluminum chloride. It has also been obtained from 3-methylcyclopentane-1,2,4-trione by catalytic hydrogenation[5] and Wolff–Kishner reduction.[6] The base-promoted cyclization of 4-oxohexanoic acid ethyl ester and diethyl propionylsuccinate with tertiary alkoxides was first reported by Bucourt.[7] The present cyclization process provides an experimentally simple route to 2-methyl-1,3-cyclopentanedione. Using the same procedure, 4-oxoheptanoic acid ethyl ester has been cyclized to give 2-ethyl-1,3-cyclopentanedione in 46% yield.

1. Chemical Research Department, Hoffmann–La Roche Inc., Nutley, New Jersey 07110.
2. For leading references, see R. T. Blickenstaff, A. C. Ghosh, and G. C. Wolf, "Total Synthesis of Steroids," Academic Press, New York, N. Y., 1974.
3. H. Schick, G. Lehmann, and G. Hilgetag, *Chem. Ber.*, **102**, 3238 (1969).
4. V. J. Grenda, G. W. Lindberg, N. L. Wendler, and S. H. Pines, *J. Org. Chem.*, **32**, 1236 (1967).
5. M. Orchin and L. W. Butz, *J. Amer. Chem. Soc.*, **65**, 2296 (1943).
6. J. P. John, S. Swaminathan, and P. S. Venkataramani, *Org. Syn.*, **47**, 83 (1967).
7. R. Bucourt, A. Pierdet, G. Costerousse, and E. Toromanoff, *Bull. Soc. Chim. Fr.*, 645 (1965).

Appendix
Chemical Abstracts Nomenclature; (Collective Index Number; Registry Numbers)

1,3-Cyclopentanedione, 2-methyl- (8,9); (765-69-5)

Succinic acid (8); Butanedioic acid (9); (110-15-6)

Propionyl chloride (8); Propanoyl chloride (9); (79-03-8)

2-Buten-2-ol, acetate (8,9); (6203-88-9)

1,2,4-Cyclopentanetrione, 3-methyl- (8,9); (4505-54-8)

Heptanoic acid, 4-oxo-, ethyl ester (8,9); (14369-94-9)

1,3-Cyclopentanedione, 2-ethyl- (8,9); (823-36-9)

MACROCYCLIC POLYAMINES:
1,4,7,10,13,16-HEXAAZACYCLOOCTADECANE

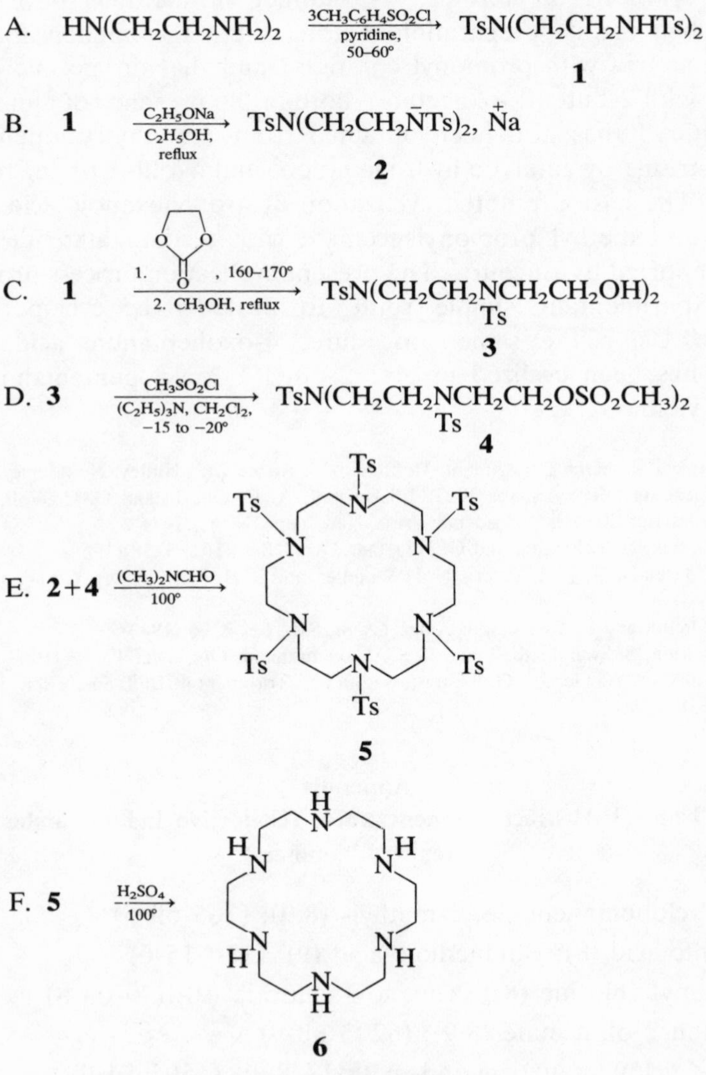

A. $HN(CH_2CH_2NH_2)_2$ $\xrightarrow[\substack{\text{pyridine,} \\ 50\text{–}60°}]{3CH_3C_6H_4SO_2Cl}$ $TsN(CH_2CH_2NHTs)_2$

$$\mathbf{1}$$

B. **1** $\xrightarrow[\substack{C_2H_5OH, \\ \text{reflux}}]{C_2H_5ONa}$ $TsN(CH_2CH_2\bar{N}Ts)_2,\ \overset{+}{Na}$

$$\mathbf{2}$$

C. **1** $\xrightarrow[\text{2. CH}_3\text{OH, reflux}]{\substack{1.\ \text{(ethylene carbonate), 160–170°}}}$ $TsN(CH_2CH_2\underset{Ts}{N}CH_2CH_2OH)_2$

$$\mathbf{3}$$

D. **3** $\xrightarrow[\substack{(C_2H_5)_3N,\ CH_2Cl_2, \\ -15\ \text{to}\ -20°}]{CH_3SO_2Cl}$ $TsN(CH_2CH_2\underset{Ts}{N}CH_2CH_2OSO_2CH_3)_2$

$$\mathbf{4}$$

E. **2+4** $\xrightarrow[100°]{(CH_3)_2NCHO}$

$$\mathbf{5}$$

F. **5** $\xrightarrow[100°]{H_2SO_4}$

$$\mathbf{6}$$

$$Ts = p\text{-}CH_3C_6H_4SO_2-$$

Submitted by T. J. Atkins, J. E. Richman, and W. F. Oettle[1]
Checked by K. Bernauer, F. Schneider, and A. Brossi

1. Procedure

A. *N,N′,N″-Tris(p-tolylsulfonyl)diethylenetriamine* (**1**).   A 5-l., three-necked, round-bottomed flask is equipped with a mechanical stirrer, reflux condenser, thermometer, and addition funnel. In the flask are placed 1150 g. (6.03 mole) of *p*-toluenesulfonyl chloride (Note 1) and 3 l. of pyridine. The mixture is stirred and warmed to 50° to dissolve the solid, the flask is immersed in a 30° water bath, and a solution of 206 g. (2.0 mole) of diethylenetriamine (Note 1) in 300 ml. of pyridine is added through the addition funnel at a rate that maintains a reaction temperature of 50–60° (1 hour). The reaction mixture is kept at 50–60° for 30 minutes longer, cooled, and divided into two equal portions in 4-l. Erlenmeyer flasks. The pyridine solutions are mechanically stirred as 1000 ml. of water is slowly poured into each. After stirring overnight and finally cooling in an ice bath for 2 hours, the white solid is collected by filtration, thoroughly washed with ice-cold 95% ethanol, and dried in a vacuum oven at 100°. The yield of triamine **1** is 950–1015 g. (84–90%), m.p. 173–175°.

B. *N,N′,N″-Tris(p-tolylsulfonyl)diethylenetriamine-N,N″-disodium salt* (**2**).   A 3-l., three-necked, round-bottomed flask is equipped with a mechanical stirrer, reflux condenser, and addition funnel. In the flask are placed 1 l. of absolute ethanol and 425 g. (0.75 mole) of triamine **1** under nitrogen. The stirred slurry is heated to reflux, the heat source is removed, and 1000 ml. of 1.5N sodium ethoxide solution (Note 2) is added through the addition funnel as rapidly as possible. The solution is then decanted from any undissolved residue into an Erlenmeyer flask. The disodium salt **2**, which crystallizes on standing overnight, is filtered under nitrogen, washed with absolute ethanol, and dried in a vacuum oven at 100°. The yield is 400–440 g. (87–96%).

C. *3,6,9-Tris(p-tolylsulfonyl)-3,6,9-triazaundecane-1,11-diol* (**3**). A 2-l., three-necked, round-bottomed flask is equipped with a mechanical stirrer, thermometer, reflux condenser, and heating mantle. In the flask are placed 226 g. (0.40 mole) of triamine **1**, 77.5 g. (0.88 mole) of ethylene carbonate (Note 1), and 0.7 g. of powdered potassium hydroxide. The stirred mixture is heated at 160–170° for 4 hours (Note 3). The reaction mixture is then allowed to cool to 90°, and 500 ml. of methanol is added through the condenser as rapidly as possible. The solution is refluxed for 30 minutes, treated with 5 g. of activated carbon, and filtered through Celite. Water (120–140 ml.) is added dropwise to the stirred filtrate until the cloud point is reached. After crystallization is complete, the diol **3** is collected and washed with 3:1 water–ethanol and dried in a vacuum oven at 50°. The yield is 225–240 g. (86–92%), of colorless product, m.p. 108–112° (Note 4).

D. *3,6,9-Tris(p-tolylsulfonyl)-3,6,9-triazaundecane-1,11-dimeth-anesulfonate* (**4**). A 3-l., three-necked, round-bottomed flask is equipped with a mechanical stirrer, addition funnel, nitrogen inlet, and low-temperature thermometer. The flask is charged with a dried solution of 200 g. (0.306 mole) of diol **3** and 100 ml. of triethylamine in 1500 ml. of methylene chloride (Note 5). The stirred solution is held at −15 to −20° in a dry-ice–acetone bath as 50 ml. (74 g.) of methanesulfonyl chloride (Note 1) is added over 10 minutes. The dry-ice bath is replaced by an ice bath, and the solution is stirred for 30 minutes, poured into a mixture of 1 l. of crushed ice and 500 ml. of 10% aqueous HCl solution, and shaken. The layers are separated, and the organic layer is washed with two 500-ml. portions of water and 500 ml. of saturated salt solution, then dried over anhydrous magnesium sulfate. The solution is filtered and evaporated to dryness under reduced pressure to give a white solid, which is dissolved in 250 ml. of methylene chloride and crystallized by addition of 500 ml. of ethyl acetate and cooling in an ice bath. The yield of methanesulfonate **4** is 215–235 g. (87–95%), m.p. 146–148°.

E. *1,4,7,10,13,16-Hexakis(p-tolylsulfonyl)-1,4,7,10,13,16-hexa-azacyclooctadecane* (**5**). A 5-l., three-necked, round-bottomed flask is equipped with a mechanical stirrer, thermometer, and an addition funnel. In the flask are placed 151 g. (0.248 mole) of sodium salt **2** and 2000 ml. of dimethylformamide. The stirred solution is held at 100° as a solution of 200 g. (0.247 mole) of sulfonate **4** in 800 ml. of dimethylformamide is added dropwise over 3 hours. After 30 minutes the heat source is removed, and 500 ml. of water is added through the addition funnel. After cooling to room temperature and stirring overnight, the cyclic hexamine **5** is collected by filtration, washed with 95% ethanol, and dried in a vacuum oven at 100°. The yield is 206–225 g. (70–77%), m.p. 260–290° (Note 6).

F. 1,4,7,10,13,16-*Hexaazacyclooctadecane* (**6**). A 3-l., three-necked, round-bottomed flask is equipped with mechanical stirrer, nitrogen inlet, and addition funnel. In the flask are placed 200 g. (0.169 mole) of cyclic hexamine **5** and 500 ml. of concentrated (97%) sulfuric acid. The stirred mixture is held at 100° for 70 hours and then cooled in ice as 1300 ml. of anhydrous ethyl ether is slowly added. The precipitated polyhydrosulfate salt is filtered under nitrogen and washed with anhydrous ethyl ether (Note 7). The salt is then stirred in 200 ml. of water and cooled in ice as 71 ml. of aqueous 50% NaOH is added to neutralize the solution. Three grams of activated carbon are added, and the solution is heated to 80° and filtered through Celite. The filtrate is cooled in ice and reacidified to pH 1 by adding 42 ml. of concentrated sulfuric acid. The white, nonhygroscopic tris(sulfuric acid) salt of **6** that precipitates is collected and washed with 95% ethanol.

To the salt and 200 ml. of water in a 1-l., round-bottomed flask equipped with an efficient magnetic stirrer and cooled in ice is added 400 ml. of 50% sodium hydroxide solution. The resulting mixture is then continuously extracted with tetrahydrofuran for 4 days (Note 8). The extract is concentrated to dryness at reduced pressure, and 1,4,7,10,13,16-hexaazacyclooctadecane (**6**) is recrystallized from acetonitrile (30 ml. per g.), giving 19–22 g. (49–50%) of long white needles, m.p. 147–150° (Note 9).

2. Notes

1. Purchased from Aldrich Chemical Company, Inc.

2. Prepared just prior to use by dissolving 34.5 g. of sodium metal in 1000 ml. of absolute ethanol under nitrogen.

3. At 100–120° the solid begins to dissolve and carbon dioxide evolution commences.

4. Pure diol **3**, m.p. 110–112°, may be obtained by recrystallization from toluene (10 ml. per g.), but further purification is unnecessary for use in Step D.

5. This solution should be dried over 4-Å molecular sieves overnight.

6. The submitters obtained a yield of 190–210 g. (65–71%), m.p. 290–315° (dec.).

7. At this stage the grayish salt is quite hygroscopic and should be carefully kept from air to prevent difficulty in filtering.

8. A large excess of base is needed to reduce the water solubility of the amine. The precipitated solids contain product; any lumps should be broken up and the aqueous slurry efficiently stirred during the extraction.

9. The submitters found m.p. 154–156.

3. Discussion

Macrocyclic polyamines and amine-ethers can be readily prepared without high-dilution techniques by this improved general procedure.[2] Previous methods employed high-dilution techniques[3-5] or transition metal templates.[6] By adapting the present procedures, macrocycles of up to 24 members may be designed and directly synthesized in high yields from readily available starting materials.[2,7-9]

The critical cyclization step gives 50–85% yields when hydrocarbon segments between heteroatoms are short (see Table I) and relatively equal segments of the target macrocycle are condensed. Methane- and p-toluenesulfonate esters[10,11] give markedly better yields than dihalides (see Table II). Cyclizations in dimethylformamide solvent are generally more convenient, although compara-

TABLE I
CONDENSATIONS OF TERMINAL AL-
KANE DITOSYLATES, $TsO–(CH_2)_n–OTs$

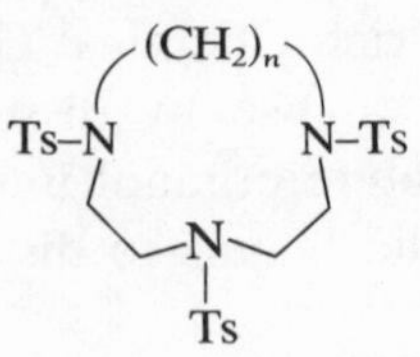

$(CH_2)_n$
Ts–N   N–Ts
N
|
Ts

n	Yield (%)
2	71
3	84
4	81
5	55
6	40–50

TABLE II

Yields of Cyclization for Various Leaving Groups

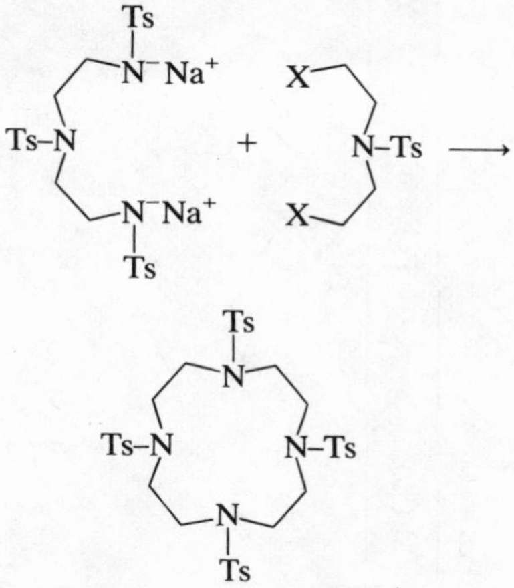

X	Yield (%)
–OTs	80
–OMs	66
–Cl	42
–Br	40
–I	25

ble yields are obtained in dimethylsulfoxide and hexamethylphosphoramide. Ether linkages or selectively substituted nitrogen may replace N–Ts groups along the chain without seriously affecting cyclization yields. Other macrocyclic amines and aminoethers[2,7–9,13] prepared by these methods are listed in Tables III and IV, respectively.

The hydroxyethylation of *sec*-sulfonamides has been adapted from Niederprüm, Voss, and Wechsberg.[12] Many new terminal diols for cyclization can be readily prepared by this method.

TABLE III
Yields of Macrocyclic Polyamines

Polyamine	Yield (%)	Polyamine	Yield (%)	Polyamine	Yield (%)
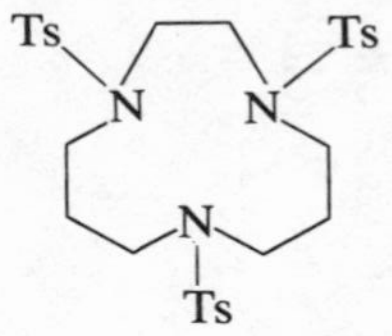 (structure)	58	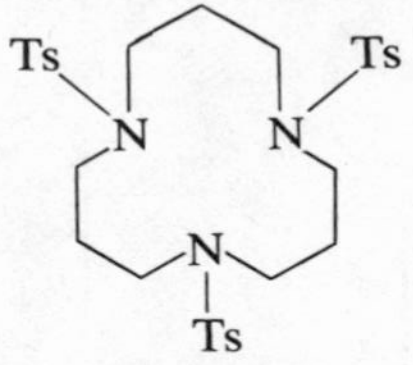 (structure)	67	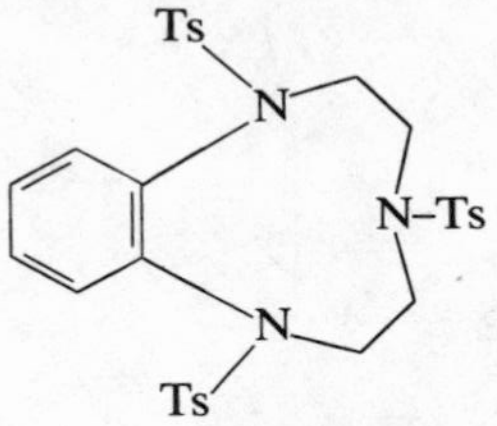 (structure)	24
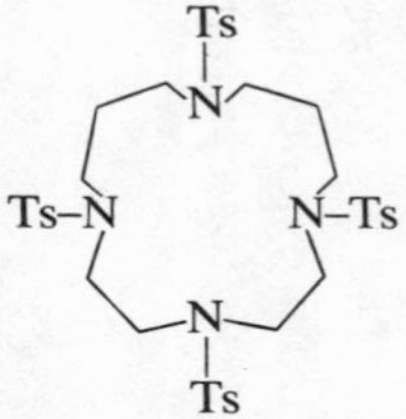 (structure)	80	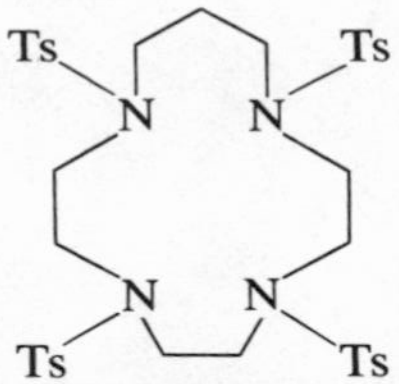 (structure)	77	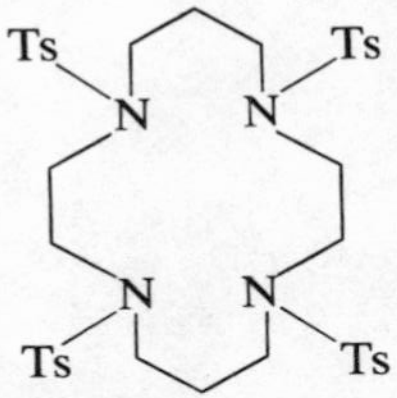 (structure)	70

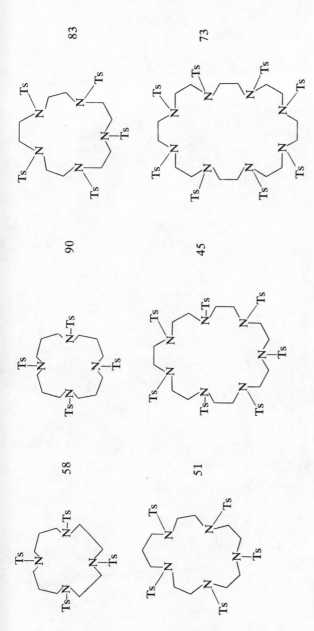

83

73

90

45

58

51

93

TABLE IV
YIELDS OF MACROCYCLIC AMINE–ETHERS

Amine–Ether	Yield (%)	Amine–Ether	Yield (%)	Amine–Ether	Yield (%)
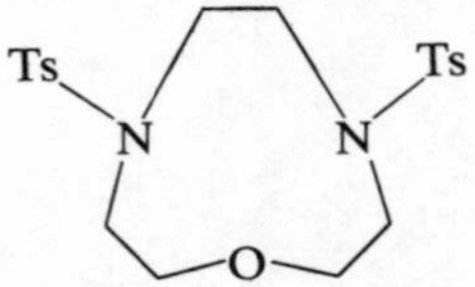 (structure)	32^b 79^c	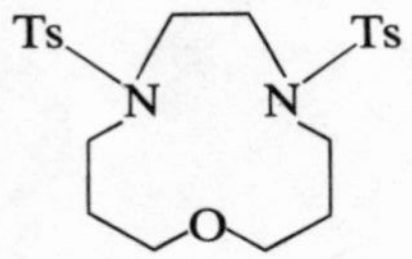 (structure)	63^b	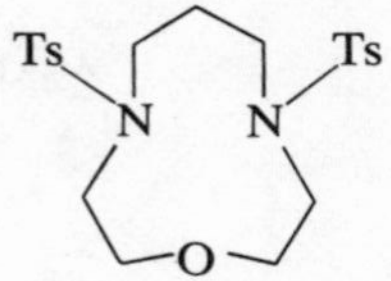 (structure)	66^b 47^c
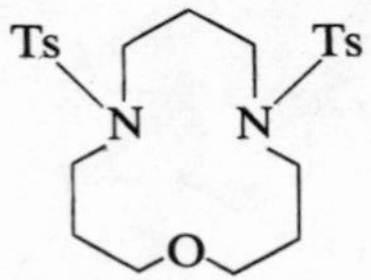 (structure)	38^b	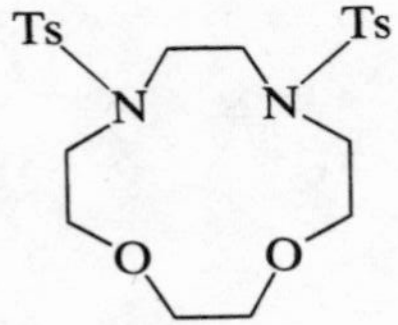 (structure)	52^b 54^c	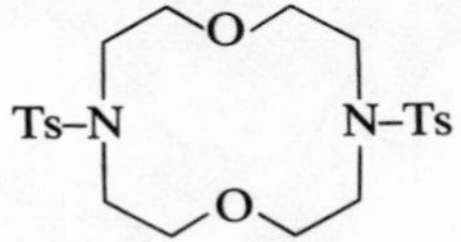 (structure)	80^a
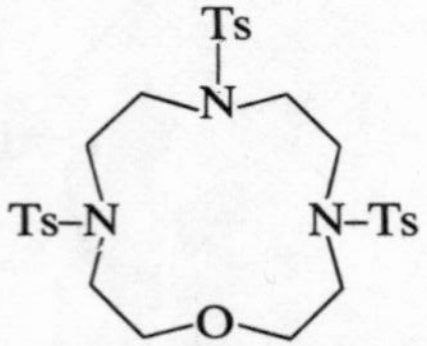 (structure)	90^a 63^b	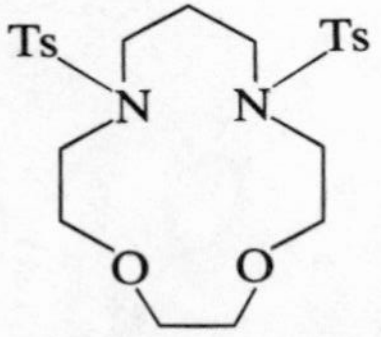 (structure)	28^b 69^c	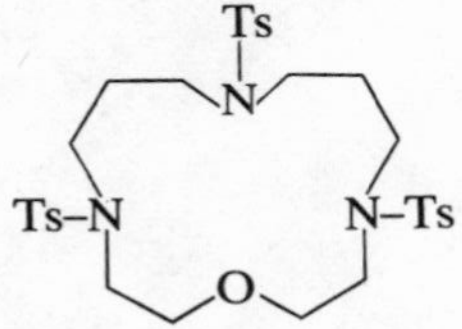 (structure)	58^c

94

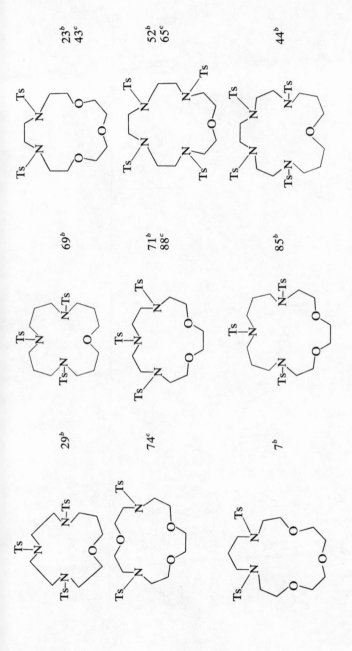

TABLE IV (Contd.)

Amine-Ether	Yield (%)	Amine-Ether	Yield (%)	Amine-Ether	Yield (%)
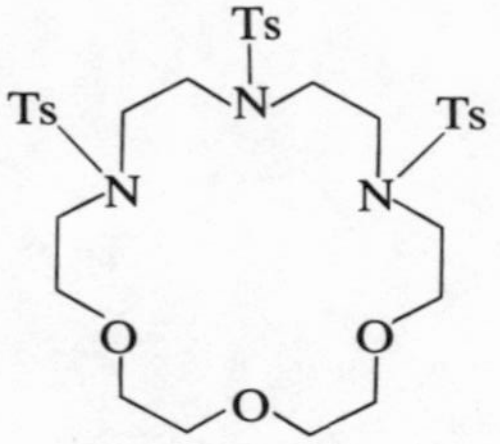	14[b] 44[c]	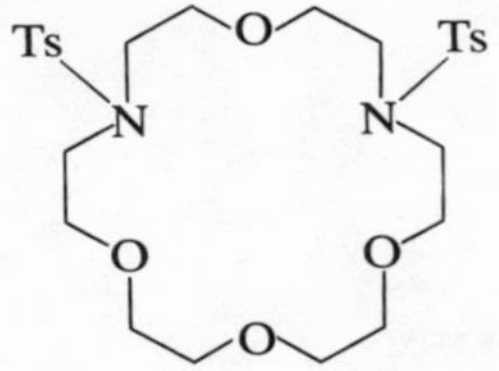	69[c]	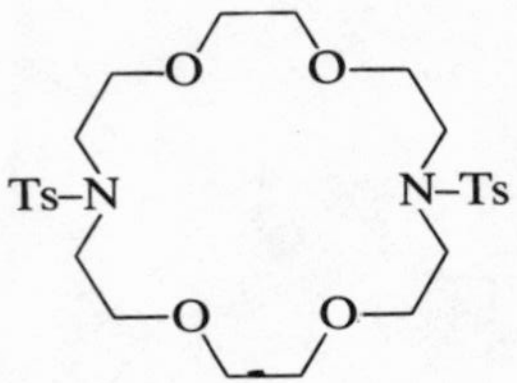	80[a]
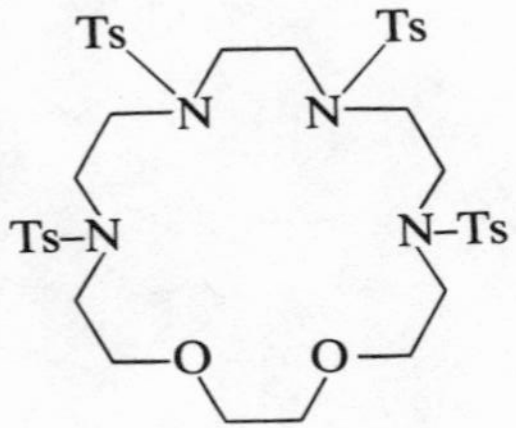	25[b] 72[a]	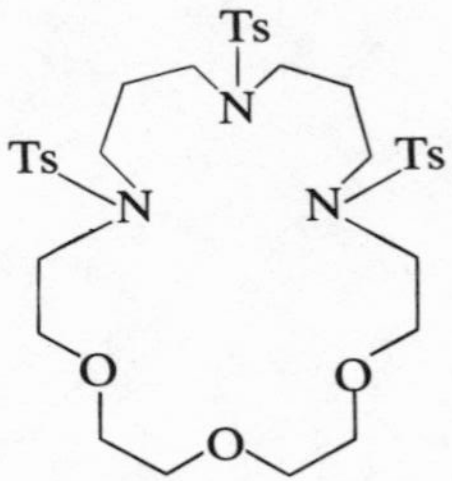	83[b]	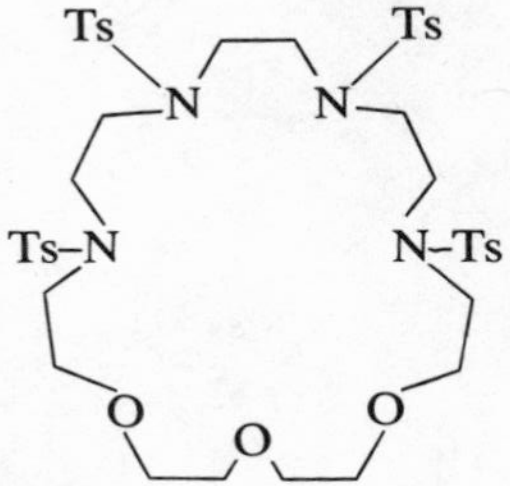	35[b]

[a] Reference 2.
[b] Reference 7.
[c] T. J. Atkins, unpublished results.

96

1. Central Research and Development Department, Experimental Station, E. I. duPont de Nemours and Company, Wilmington, Delaware 19898.
2. J. E. Richman and T. J. Atkins, *J. Amer. Chem. Soc.*, **96**, 2268 (1974).
3. (a) H. Stetter and E.-E. Roos, *Chem. Ber.*, **87**, 566 (1954); (b) H. Stetter and J. Marx, *Justus Liebigs Ann. Chem.*, **607**, 59 (1957); (c) H. Stetter and K.-H. Mayer, *Chem. Ber.*, **94**, 1410 (1961).
4. H. E. Simmons and C. H. Park, *J. Amer. Chem. Soc.*, **90**, 2428 (1968).
5. B. Dietrich, J. M. Lehn, J. P. Sauvage, and J. Blanzat, *Tetrahedron*, 1629 (1973).
6. N. F. Curtis, *Coord. Chem. Rev.*, **3**, 3 (1968).
7. W. Rasshofer, W. Wehner, and F. Vögtle, *Justus Liebigs Ann. Chem.*, 916 (1976).
8. W. Rasshofer and F. Vögtle, *Justus Liebigs Ann. Chem.*, 1340 (1977).
9. E. Buhleier, W. Rasshofer, W. Wehner, F. Luppertz, and F. Vögtle, *Justus Liebigs Ann. Chem.*, 1344 (1977).
10. The preparation of the dimesylate in the procedure is essentially that of R. K. Crossland and K. L. Servis, *J. Org. Chem.*, **35**, 3195 (1970).
11. Ditosylates can be prepared by the procedure described in C. S. Marvel and V. C. Sekera, *Org. Syn.*, Coll. Vol. **3**, 366 (1955).
12. H. Niederprüm, P. Voss, and M. Wechsberg, *Justus Liebigs Ann. Chem.*, 11 (1973).
13. T. J. Atkins, J. E. Richman, and W. F. Oettle, unpublished results.

Appendix
Chemical Abstracts Nomenclature; (Collective Index Number; Registry Numbers)

1,4,7,10,13,16-Hexaazacyclooctadecane (8,9); (296-35-5)

Diethylenetriamine, N,N',N''-tris(p-tolylsulfonyl)- (8); Benzenesulfonamide, 4-methyl-N,N-bis[2-[[(4-methylphenyl)sulfonyl]amino]-ethyl]- (9); (56187-04-3)

p-Toluenesulfonyl chloride (8); Benzenesulfonyl chloride, 4-methyl- (9); (98-59-9)

Pyridine (8,9); (110-86-1)

Diethylenetriamine (8); 1,2-Ethanediamine, N-(2-aminoethyl)- (9); (111-40-0)

N,N',N''-Tris(p-tolylsulfonyl)diethylenetriamine-N,N''-disodium salt: Diethylenetriamine, N,N',N''-tris(p-tolylsulfonyl)-, disodium salt (8); Benzenesulfonamide, 4-methyl-N,N-bis[[[(4-methylphenyl)sulfonyl]amino]ethyl]-, disodium salt (9); (52601-80-6)

Sodium ethoxide: Ethyl alcohol, sodium salt (8); Ethanol, sodium salt (9); (141-52-6)

3,6,9-Tris(p-tolylsulfonyl)-3,6,9-triazaundecane-1,11-diol; ($-$)

Ethylene carbonate: 1,3-Dioxolan-2-one (8,9); (96-49-1)

3,6,9-Tris(p-tolylsulfonyl)-3,6,9-triazaundecane-1,11-dimethanesulfonate; (−)

Triethylamine (8); Ethanamine, N,N-diethyl- (9); (121-44-8)

Methylene chloride: Methane, dichloro- (8,9); (75-09-2)

Methanesulfonyl chloride (8,9); (124-63-0)

Ethyl acetate: Acetic acid, ethyl ester (8,9); (141-78-6)

1,4,7,10,13,16-Hexaazacyclooctadecane, 1,4,7,10,13,16-hexakis-(p-tolylsulfonyl)- (8); 1,4,7,10,13,16-Hexaazacyclooctadecane, 1,4,7,10,13,16-hexakis[(methylphenyl)sulfonyl]- (9); (52601-75-9)

Dimethylformamide: Formamide, N,N-dimethyl- (8,9); (68-12-2)

Tetrahydrofuran: Furan, tetrahydro- (8,9); (109-99-9)

Toluene (8); Benzene, methyl- (9); (108-88-3)

Dimethylsulfoxide: Methyl sulfoxide (8); Methane, sulfinylbis- (9); (67-68-5)

Hexamethylphosphoramide: Phosphoric triamide, hexamethyl- (8,9); (680-31-9)

MACROLIDES FROM CYCLIZATION OF ω-BROMOCARBOXYLIC ACIDS: 11-HYDROXYUNDECANOIC LACTONE
(Oxacyclododecan-2-one)

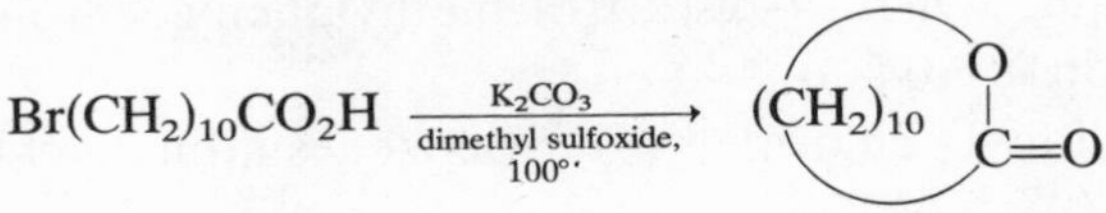

$$Br(CH_2)_{10}CO_2H \xrightarrow[\substack{\text{dimethyl sulfoxide,} \\ 100°}]{K_2CO_3} (CH_2)_{10}\begin{array}{c}O\\ \\ C=O\end{array}$$

Submitted by C. GALLI and L. MANDOLINI[1]
Checked by KAORU MORI and CARL R. JOHNSON

1. Procedure

A 1-l., three-necked, round-bottomed flask is equipped with an internal thermometer, mechanical stirrer, dropping funnel, and calcium chloride drying tube. The flask is charged with 500 ml. of dimethyl sulfoxide and 15 g. (0.11 mole) of potassium carbonate (Note 1). The mixture is heated to 100°, and a solution of 10.0 g. (0.038 mole) of 11-bromoundecanoic acid (Note 2) in 200 ml. of dimethyl sulfoxide is added dropwise with vigorous stirring over 1 hour. After cooling at room temperature, the mixture is decanted and filtered free of any suspended solid material (Note 3) through a Büchner funnel with occasional suction. The solid residue is rinsed with 50 ml. of dimethyl sulfoxide, and the washings are added to the original filtrate. The resulting clear solution is diluted with 250 ml. of water and extracted with three 250-ml. portions of petroleum ether. The combined organic layers are washed with 200 ml. of water, dried over anhydrous sodium sulfate, and concentrated to obtain *ca.* 7 g. of crude material. A simple distillation at reduced pressure from a small Claisen flask yields 5.5–5.8 g. (79–83%) of pure 11-hydroxyundecanoic lactone as a colorless, musk-smelling liquid, b.p. 124–126° (13 mm.), n^{19}D 1.4721 (Notes 4 and 5). The residue is ground with 5 ml. of hexane and filtered to afford 0.4–0.7 g. (6–10%) of the 24-membered dilactone 1,13-dioxacyclotetracosane-2,14-dione as white crystals, m.p. 71.5–72° (from hexane) (Note 6).

2. Notes

1. Reagent-grade dimethyl sulfoxide and anhydrous potassium carbonate were used.

2. 11-Bromoundecanoic acid, available from Aldrich Chemical Company, Inc., was used without further purification.

3. Filtration is optional. However, it does reduce the extent of emulsion formation during the subsequent extractions.

4. The submitters report that the pure lactone and dilactone can also be obtained by chromatography of the crude product on silica gel with chloroform as the eluant.

5. The product is pure by vapor phase and thin-layer chromatography; infrared (carbon tetrachloride) cm^{-1}: 1740; proton magnetic resonance (carbon tetrachloride) δ (multiplicity, number of protons, assignment): 4.14 (broad triplet, 2, CH_2O), 2.30 (broad triplet, 2, CH_2CO), and 1.8–1.2 (multiplet, 16).

6. Stoll and Rouvè[2] report m.p. 71.5–72°.

3. Discussion

Available methods for the synthesis of macrolides involve the cyclization of long-chain bifunctional precursors,[3] depolymerization processes,[4] ring-enlargement reactions,[5] and special methods such as the thermal decomposition of tricycloalkylidene peroxides.[6] The method reported here is essentially that of the submitters.[7] Its improvements result from a quantitative approach to the cyclization of a series of ω-bromo fatty acids under conditions well defined from the kinetic point of view. A unique feature of this procedure in comparison with other methods involving cyclization of α,ω-bifunctional precursors, which are generally run under Ziegler's high-dilution conditions, is that high rates of feed can be used, so that the special devices usually employed for the slow addition of the reagent into the reaction medium are not required. The synthesis is characterized by relatively mild reaction conditions and simple work-up. Moreover, it is suited for relatively large-scale preparations. Up to 50 g. of 11-bromoundecanoic acid can be cyclized in more than 70% yield in a single run, employing no more than 1 l. of solvent and an addition time of 3–4 hours.

The reaction illustrates a typical example of the preparation of macrolides. Lactones with more than 12 members can be obtained in even better yields. For example, 15-hydroxypentadecanoic lactone (m.p. 35–37°) and 17-hydroxyheptadecanoic lactone (m.p. 40–41°) were prepared by the submitters in about 95% yield in a practically pure form, with no trace of the corresponding dilactones.

Recent progress in chemistry and biochemistry of macrolides was recently reviewed.[8]

1. Centro di Studio sui Meccanismi di Reazione del Consiglio Nazionale delle Ricerche, c/o Istituto Chimico dell' Università di Roma, 00185, Rome, Italy.
2. M. Stoll and A. Rouvè, *Helv. Chim. Acta*, **18**, 1119 (1935).
3. E. J. Corey and K. C. Nicolau, *J. Amer. Chem. Soc.*, **96**, 5614 (1974), and references therein.
4. E. W. Spanagel and W. H. Carothers, *J. Amer. Chem. Soc.*, **58**, 654 (1936).
5. I. J. Borowitz, V. Bandurco, M. Heyman, R. D. G. Rigby, and S. Ueng, *J. Org. Chem.*, **38**, 1234 (1973); C. H. Hassall, *Org. React.*, **9**, 73 (1957).
6. P. R. Story and P. Busch, *Advan. Org. Chem.*, **8**, 67 (1972).
7. C. Galli and L. Mandolini, *Gazz. Chim. Ital.*, **105**, 367 (1975).
8. S. Masamune, G. S. Bates, and J. W. Corcoran, *Angew. Chem. Int. Ed. Engl.*, **16**, 585 (1977).

Appendix
Chemical Abstracts Nomenclature (Collective Index Number; Registry Numbers)

Undecanoic acid, 11-hydroxy-, lactone (8); Oxacyclododecan-2-one (9); (1725-03-7)

Dimethyl sulfoxide: Methyl sulfoxide (8); Methane, sulfinylbis- (9); (67-68-5)

Potassium carbonate: Carbonic acid, dipotassium salt (8,9); (584-08-7)

Undecanoic acid, 11-bromo- (8,9); (2834-05-1)

Hexane (8,9); (110-54-3)

1,3-Dioxacyclotetracosane-2,14-dione; (−)

15-Hydroxypentadecanoic lactone: Oxacyclohexadecan-2-one (8,9); (106-02-5)

17-Hydroxyheptadecanoic lactone: Oxacyclooctadecan-2-one (8,9); (5637-97-8)

NITRILES FROM KETONES: CYCLOHEXANECARBONITRILE

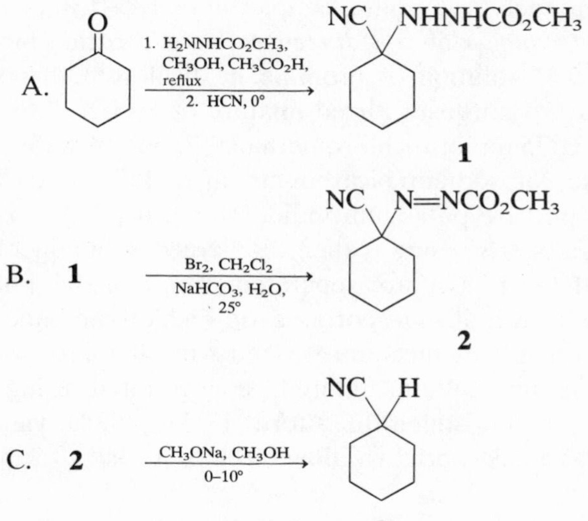

Submitted by P. A. WENDER,[1] M. A. EISSENSTAT,[1] N. SAPUPPO,[1] and F. E. ZIEGLER[2]
Checked by D. F. BUSHEY and G. BÜCHI

1. Procedure

Caution! Because of the toxicity of hydrogen cyanide, this procedure should be conducted in a well-ventilated hood and rubber gloves should be worn. (See also Note 3.)

A. 2-(1-*Cyanocyclohexyl*)*hydrazinecarboxylic acid methyl ester* (**1**). A 100-ml., three-necked flask is equipped with a reflux condenser, thermometer, magnetic stirring bar, and gas-exhaust tube (Note 1). The flask is charged with 9.0 g. (0.10 mole) of methyl carbazate (Note 2), 20 ml. of methanol, 2 drops of acetic acid, and 9.8 g. (0.10 mole) of cyclohexanone. The resulting mixture is refluxed for 30 minutes, then cooled to 0° and treated with 6 ml. (0.15 mole) of hydrogen cyanide (Note 3), added dropwise over a period of 3 minutes. After approximately 15 minutes the solution is allowed to warm to room temperature, during which time the hydrazine **1** crystallizes (Notes 4 and 5). After 2 hours the resulting mixture is vacuum filtered, and the crystalline residue is washed with 10 ml. of cold methanol (Note 6) to afford 17.7 g. of hydrazine **1**. Concentration of the filtrate provides an additional 1.4 g. of product. The total yield of crude hydrazine **1**, m.p. 130–133°, (Note 7) is 97%. Recrystallization (methanol–pentane) provides an analytically pure sample, m.p. 135–136°.

B. 2-(1-*Cyanocyclohexyl*)*diazenecarboxylic acid methyl ester* (**2**). A 5.5 M solution of bromine in dichloromethane is added dropwise to a vigorously stirred mixture of 19.1 g. (0.097 mole) of hydrazine **1**, 75 ml. of dichloromethane, 75 ml. of water, and 22 g. (0.262 mole) of sodium bicarbonate in a 250-ml. flask, until a persistent, positive potassium iodide–starch paper test is obtained (Note 8). Excess bromine is then discharged with aqueous sodium sulfite, and the phases are separated. The aqueous phase is extracted with two 50-ml. portions of dichloromethane, and the combined organic extracts are washed with 30 ml. of water, dried over magnesium sulfate, filtered, concentrated using a rotary evaporator, and distilled to afford 17.3 g. (93% yield) of the diazene **2** as a clear, bright yellow oil, b.p. 95–97° (0.2 mm.) (Note 7).

C. *Cyclohexanecarbonitrile* (**3**). A 100-ml., three-necked, round-bottomed flask is fitted with a thermometer, a magnetic stirring bar, and an addition funnel. The flask is charged with 2.7 g. (0.050 mole) of sodium methoxide and 25 ml. of methanol. The solution is cooled in ice and stirred while a solution of 17.3 g. (0.089 mole) of diazene **2** in 10 ml. of methanol is added dropwise at a rate that maintains the solution temperature at 0–10° (Note 9). Stirring is continued for an additional 30 minutes at ambient temperature, and the mixture is then poured into 70 ml. of water. The resulting solution is extracted with five 20-ml. portions of pentane, and the combined organic extracts are dried over magnesium sulfate, filtered, concentrated, and distilled to provide 8.6–9.4 g. (78–86%) of carbonitrile **3**, b.p. 117–119° (90 mm.) (Note 10).

2. Notes

1. The tube is connected to the top of the reflux condenser and attached to a length of tygon tubing. The end of the tubing is positioned in the exhaust vent of the hood to remove any hydrogen cyanide vapor.

2. Methyl carbazate was prepared by the method of O. Diels.[3] This reagent is available from Aldrich Chemical Company, Inc., under the name methyl hydrazinocarboxylate. All other chemicals used in this sequence were of the highest purity commercially available and were not further purified before use.

3. Hydrogen cyanide can be purchased from Fumico Inc., Amarillo, Texas. When large amounts of HCN are used, it is recommended that amyl nitrite pearls and an oxygen cylinder with mask be available. These in combination are effective antidotes for HCN poisoning. The checkers prepared HCN according to the method of K. Ziegler.[4] The use of smaller quantities of HCN results in slower reaction and reduced yield.

4. Spontaneous crystallization usually occurs within 30 minutes. If it does not, crystallization is induced by scratching the bottom of the reaction flask.

5. The solid formed at this point usually interferes with stirring. The yield of product, however, is not affected by the nature of mixing beyond this point.

6. The methanol is used to transfer any residual material from the reaction flask as well as to wash the residue.

7. Further purification of this intermediate is unnecessary for the preparation of nitriles.

8. The endpoint of this oxidation is indicated by the appearance in the reaction mixture of a red-orange color caused by excess bromine.

9. Nitrogen is vigorously and instantaneously evolved as each drop is added.

10. Physical and analytical characterizations of cyclohexanecarbonitrile (3) agree with literature reports[5]: proton magnetic resonance (chloroform-d) δ (multiplicity, number of protons, assignment): 2.90–2.40 (multiplet, 1, α-H), 2.00–1.30 (multiplet, 10, CH_2). The average overall yield of cyclohexanecarbonitrile (3) from cyclohexanone is 85% when the diazene 2 is distilled and 86–90% when crude diazene is used directly. The submitter reported 8.9 g (92%) of carbonitrile 3.

3. Discussion

The transformation of ketones to nitriles has been accomplished with varying success using several methods including cyanide displacement of acetates[6] and halides[7] obtained from ketones; cyanohydrin formation, dehydration, and reduction of the unsaturated nitriles[8]; reaction of tosylhydrazones with potassium cyanide followed by pyrolysis of the cyanohydrazine[9]; and reaction of tosylmethylisocyanide with ketones.[10] Related transformations of

TABLE I
KETONE TO NITRILE TRANSFORMATION

Ketone	Diazene Yield (%)[a]	Nitrile Yield (%)[b] (ratio of diastereomers)
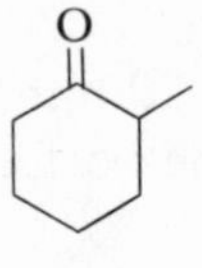 2-methylcyclohexanone (structure)	95	89 (50:50)
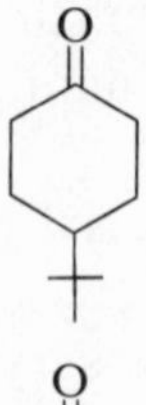 4-tert-butylcyclohexanone (structure)	94	97 (58:42)
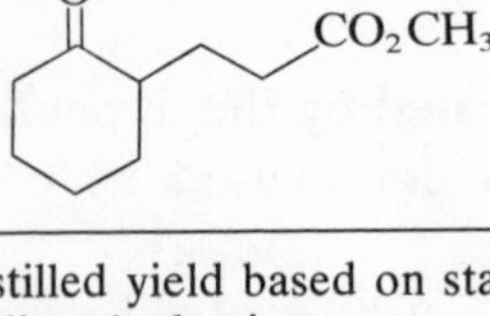 2-(methyl propanoate)cyclohexanone, CO_2CH_3 (structure)	94	90 (50:50)

[a] Distilled yield based on starting ketone, obtained without purification of intermediate hydrazine.
[b] Yield based on purified diazene as determined by chromatographic analysis.

ketones and ketone derivatives to nitrile derivatives have also been reported.[11]

This procedure uses readily available reagents and provides a simple and efficient method for nitrile synthesis. The entire sequence of four steps can be performed in a single day. Although product formation in the second step is presumably thermodynamically controlled, the cyanohydrazine is favored in all cases studied except with aryl ketones. A water–methanol solution of ammonium chloride and potassium cyanide can also be employed for the cyanohydrazine formation, but lower yields (*ca.* 60%) are obtained. The third step, a conveniently performed titration procedure with bromine as oxidant, can be effected with other oxidizing reagents such as 4-phenyl-4*H*-1,2,4-triazole-3,5-dione, *tert*-butyl hypochlorite, and Jones reagent.[12] The final diazene-decomposition step is induced with bases and nucleophiles such as methoxide, ethoxide, hydroxide, and iodide. The diazine decomposition can be extended to the preparation of α-methylnitriles and α-carboalkoxynitriles.[13]

This procedure is general for the preparation of secondary nitriles (Table I), and can be used in the presence of other functional groups by the appropriate choice of oxidation and decomposition reagents.[14]

1. Department of Chemistry, Harvard University, Cambridge, Massachusetts 02138.
2. Department of Chemistry, Yale University, New Haven, Connecticut 06520.
3. O. Diels, *Chem. Ber.*, **47**, 2183 (1914).
4. K. Ziegler, *Org. Syn.*, Coll. Vol. **1**, 314 (1941).
5. "Dictionary of Organic Compounds," Vol. 2, Oxford University Press, New York, N.Y., 1965, p. 782.
6. B. A. Dadson and J. Harley-Mason, *Chem. Commun.*, 665 (1969).
7. L. Friedman and H. Shechter, *J. Org. Chem.*, **25**, 877 (1960).
8. K. Meyer, *Helv. Chim. Acta*, **29**, 1580 (1946); N. Danieli, Y. Mazur, and F. Sondheimer, *J. Amer. Chem. Soc.*, **84**, 875 (1962).
9. S. Cacchi, L. Caglioti, and G. Paolucci, *Chem. Ind. London*, 213 (1972).
10. O. H. Oldenziel and A. M. van Leusen, *Synth. Commun.*, **2**, 281 (1972); O. H. Oldenziel, D. Van Leusen, and A. M. Van Leusen, *J. Org. Chem.*, **42**, 3114 (1977); O. H. Oldenziel, J. Wildeman, and A. M. Van Leusen, *Org. Syn.*, **57**, 8 (1977).
11. B. M. Trost, M. J. Bogdanowicz, and (in part) J. Kern, *J. Amer. Chem. Soc.*, **97**, 2218 (1975), and references contained therein.
12. L. F. Fieser and M. Fieser, "Reagents for Organic Synthesis," John Wiley & Sons, New York, N. Y., 1967, p. 142.
13. F. E. Ziegler and P. A. Wender, *J. Amer. Chem. Soc.*, **93**, 4318 (1971).
14. K. C. Mattes, M. T. Hsia, C. R. Hutchinson, and S. A. Sisk, *Tetrahedron Lett.*, 3541 (1977).

Appendix
Chemical Abstracts Nomenclature (Collective Index Number; Registry Numbers)

Cyclohexanecarbonitrile (8,9); (766-05-2)

Hydrogen cyanide: Hydrocyanic acid (8,9); (74-90-8)

Carbazic acid, 2-(1-cyanocyclohexyl)-, methyl ester (8); Hydrazinecarboxylic acid, 2-(1-cyanocyclohexyl)-, methyl ester (9); (−)

Methyl carbazate: Carbazic acid, methyl ester (8); Hydrazinecarboxylic acid, methyl ester (9); (6294-89-9)

Cyclohexanone (8,9); (108-94-1)

Formic acid, [(1-cyanocyclohexyl)azo]-, methyl ester (8); Diazenecarboxylic acid, 2-(1-cyanocyclohexyl)-, methyl ester (9); (33670-04-1)

Sodium methoxide: Methanol, sodium salt (8,9); (124-41-4)

Tosylmethylisocyanide: Methyl isocyanide, p-tolylsulfonyl- (8); Benzene, 1-[(isocyanomethyl)sulfonyl]-4-methyl- (9); (36635-61-7)

4H-1,2,4-Triazole-3,5-dione, 4-phenyl- (8,9); (4233-33-4)

tert-Butyl hypochlorite: Hypochlorous acid, tert-butyl ester (8); Hypochlorous acid, 1,1-dimethylethyl ester (9); (507-40-4)

NITRONES FOR INTRAMOLECULAR 1,3-DIPOLAR CYCLOADDITIONS: HEXAHYDRO-1,3,3,6-TETRAMETHYL-2,1-BENZISOXAZOLINE

(2,1-Benzisoxazole, 1,3,3a,4,5,6,7,7a-octahydro-1,3,3,6-tetramethyl-)

$$(CH_3)_2C{=}CHCH_2CH_2CH(CH_3)CH_2CHO + CH_3NHOH \cdot HCl \xrightarrow[\text{reflux}]{\text{NaOCH}_3, \text{ toluene,}}$$

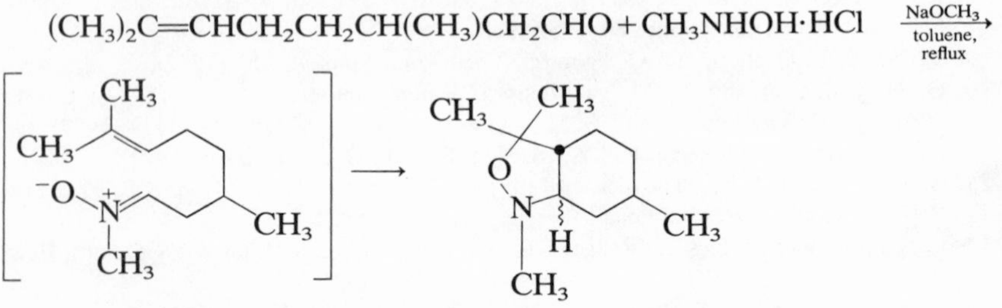

Submitted by Norman A. LeBel and Dorothy Hwang[1]
Checked by Christopher K. VanCantfort and Robert M. Coates

1. Procedure

A 1-l., three-necked, round-bottomed flask is fitted with a mechanical stirrer, a reflux condenser attached to a Dean-Stark water separator, and a 250-ml. dropping funnel. The flask is charged with 25.0 g. (0.16 mole) of 3,7-dimethyl-6-octenal (Note 1) and 500 ml. of toluene. The solution is heated to reflux with stirring, and a solution of N-methylhydroxylamine, methanol, and toluene is added (see below).

To a cooled and magnetically stirred solution of 23.4 g. (0.28 mole) of N-methylhydroxylamine hydrochloride (Notes 2 and 3) in 40 ml. of methanol is added 15.3 g. (0.282 mole) of sodium methoxide. The cooling bath is removed, and the mixture is stirred at room temperature for 15 minutes. The mixture is filtered rapidly through a 35-mm., coarse, sintered-glass funnel, and the filter cake is washed with 10 ml. of methanol. The filtrates are combined, refiltered, and mixed with 150 ml. of toluene.

The two-phase mixture containing N-methylhydroxylamine is added dropwise to the refluxing toluene solution of the aldehyde over 3 hours. During this time the distillate is collected and discarded in 25-ml. portions until the last such portion collected and discarded is clear (Note 4). Reflux and stirring are continued for an additional 3 hours, and then the clear reaction mixture is allowed to cool. The product is extracted with three 80-ml. portions of 10% aqueous hydrochloric acid. The extracts are combined, and the pH of the solution is adjusted to >12 by the slow addition of 30% aqueous potassium hydroxide. The basic mixture is extracted with two 120-ml. portions of pentane (Note 5), and the combined extracts are washed once with 100 ml. of water and dried over anhydrous potassium carbonate. The pentane is removed on a rotary evaporator and the residue is distilled through a short Vigreux column under reduced pressure to give 19.1–19.6 g. (64–67%) (Note 6) of hexahydro-1,3,3,6-tetramethyl-2,1-benz-isoxazoline, b.p. 90–92° (9 mm.) (Note 7).

2. Notes

1. Matheson Coleman and Bell technical-grade 3,7-dimethyl-6-octenal (citronellal), b.p. 87–90° (10 mm.), is used after a simple distillation.

2. This quantity amounts to an 0.72 molar excess. A molar excess of at least 0.5 is desirable to maximize the yield.

3. N-Methylhydroxylamine hydrochloride, m.p. 83–84°, purchased from Aldrich Chemical Company, Inc., was used directly. Alternatively, the hydrochloride can be prepared by the reduction of nitromethane with zinc dust and ammonium chloride.[2]

4. Water and methanol are removed by this procedure so that a higher reaction temperature can be achieved.

5. One to four additional extractions improve the yield slightly.

6. The yield is lowered by the presence of 3,7-dimethyl-7-octenal in the technical-grade 3,7-dimethyl-6-octenal (citronellal) used.

7. A gas chromatographic analysis carried out by the submitters using a 1.85 m. × 0.32 cm. stainless-steel column packed with 10% Polyglycol E-20M on Chromosorb W at 150° indicated that the product is a mixture of trans, trans and cis, trans stereoisomers in a ratio of 89 : 11. The spectral properties of the product are: infrared (thin film) cm^{-1}: 1462, 1387, 1348, 1287, 1277, 1193, 1179, 1136, 1124, 935, 915, 877, 810; proton magnetic resonance (chloroform-d) δ (multiplicity, number of protons, assignment, coupling constant J in Hz.): 1.00 (doublet, 3, CH_3, J = 7), 1.08 (singlet, 3, CH_3), 1.28 (singlet, 3, CH_3), 2.60 (singlet, 3, $N-CH_3$).

3. Discussion

This procedure is an adaptation[3] of the original method,[4] and avoids the isolation and purification of N-methylhydroxylamine. Nitrones undergo 1,3-dipolar cycloadditions with a wide variety of dipolarophiles (see recent review[5]). The intramolecular variation represents a useful synthetic approach, as carbocyclic or heterocyclic rings are generated together with the five-membered isoxazolidines.[6] The intermediate nitrone is usually not preformed (present example), although intermolecular cycloaddition is rarely a problem. N-alkyl-, N-alkenyl-, and N-arylhydroxylamines have been used with aldehydes and ketones to generate the nitrones in situ, and some typical examples are listed in Table I. Cyclic azomethine imine oxides are also important substrates for intramolecular cycloadditions.[6]

Although isoxazolidines are less basic than the analogous

TABLE I
INTRAMOLECULAR 1,3-DIPOLAR CYCLOADDITIONS OF NITRONES TO ALKENES

Carbonyl Compound	Hydroxylamine	Product	Yield (%)
$CH_2=CH(CH_2)_3CHO$	CH_3NHOH	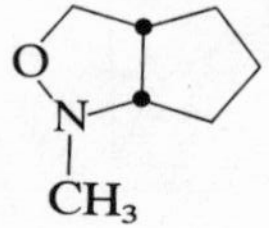	41[4]
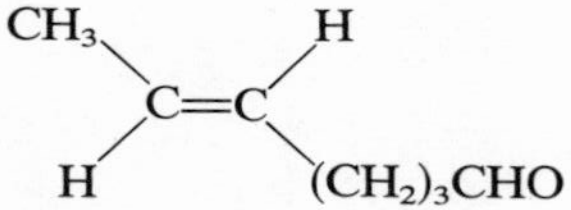	C_2H_5NHOH	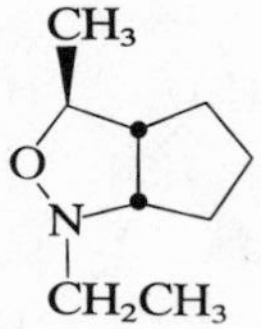	77[4]
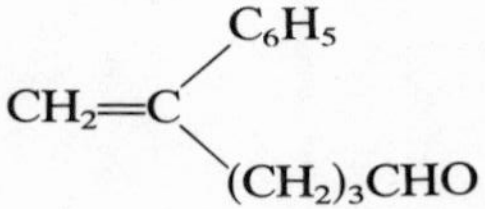	CH_3NHOH	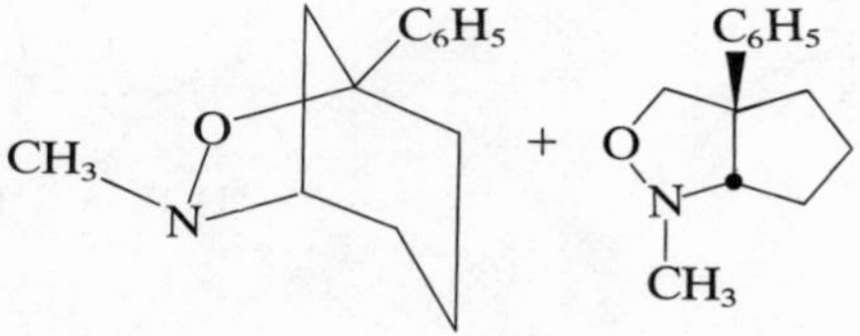 9:1 ratio	87[7]
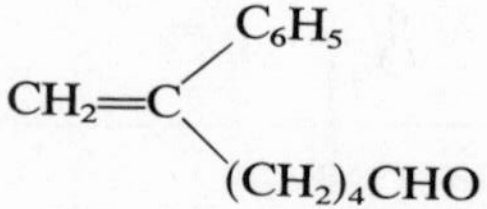	CH_3NHOH	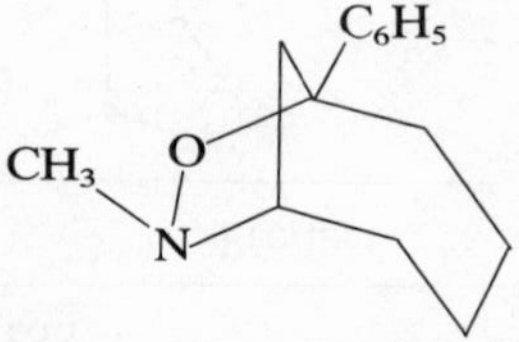	78[7]

continued overleaf

109

TABLE I (cont.)

Carbonyl Compound	Hydroxylamine	Product	Yield (%)
$CH_2=CH(CH_2)_3\overset{O}{\underset{\|\|}{C}}CH_3$	CH_3NHOH		80[8]
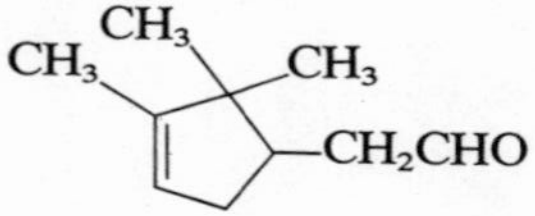 (cyclopentene with CH_3, CH_3, CH_2CHO)	$(CH_3)_2CHNHOH$	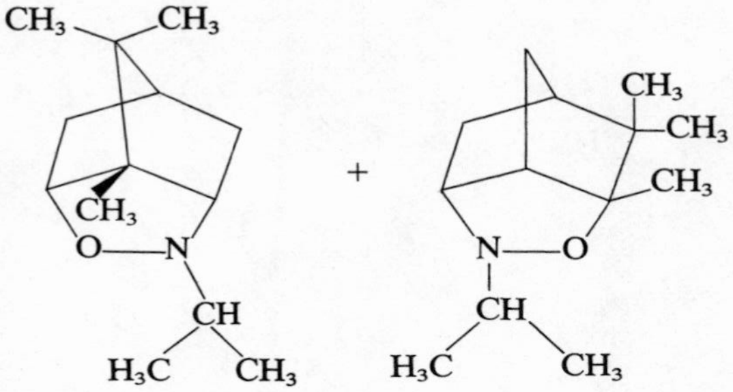	70–80[9]
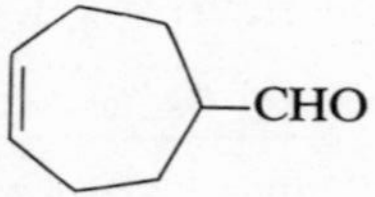 (cycloheptene-CHO)	C_6H_5NHOH	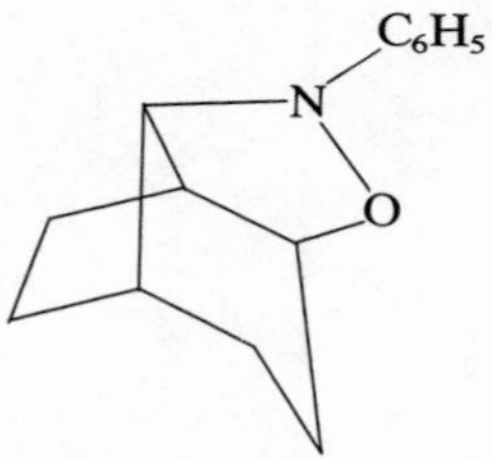	88[10]

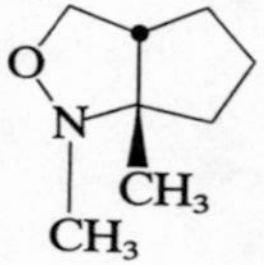

110

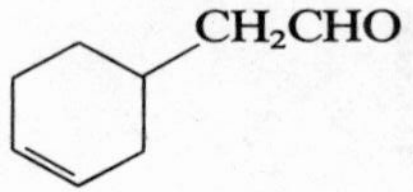

55.4^{11}

80^{12}

10^{13}

CH₃NHOH (appears three times as reagent)

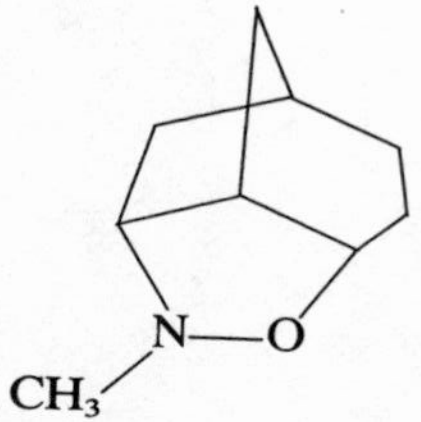

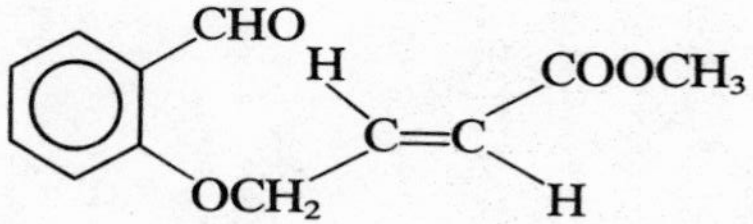

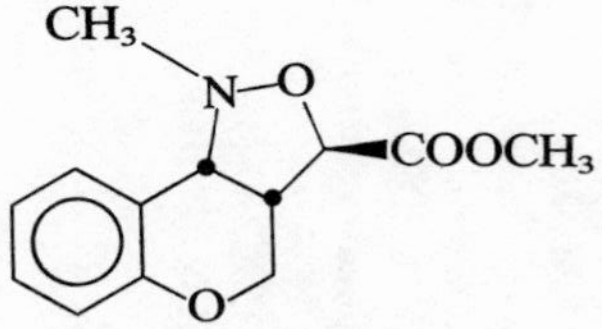

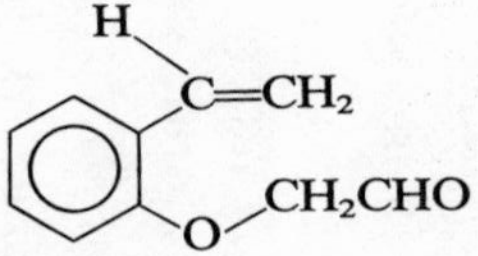

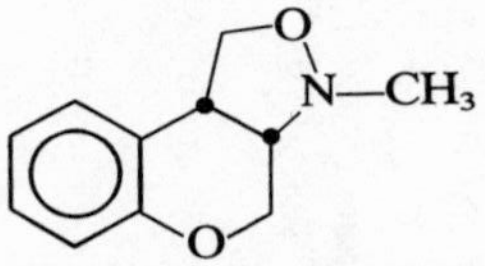

111

amines, N-alkylisoxazolidines can form quaternary ammonium salts. Reductive cleavage of isoxazolidines and the methiodides can be effected with various reagents (zinc–acetic acid, hydrogen–palladium, lithium aluminum hydride), and the yields of 1,3-aminoalcohols are generally excellent.[4] Other reagents that result in modifications of the isoxazolidine ring include peroxyacids,[14] strong bases,[15] triplet photosensitizers,[15] and cyanogen bromide.[7]

1. Department of Chemistry, Wayne State University, Detroit, Michigan 48202.
2. E. Beckmann, *Justus Liebigs Ann. Chem.*, **365**, 201 (1909).
3. N. A. LeBel and E. G. Banucci, *J. Org. Chem.*, **36**, 2440 (1971).
4. N. A. LeBel, M. E. Post, and J. J. Whang, *J. Amer. Chem. Soc.*, **86**, 3759 (1964).
5. D. St. C. Black, R. F. Crozier, and V. C. Davis, *Synthesis*, 205, (1975).
6. For a survey of intramolecular 1,3-dipolar cycloadditions, see A. Padwa, *Angew. Chem. Int. Ed. Engl.*, **15**, 123 (1976).
7. R. J. Newland, *Diss. Abstr. Int. B.*, **35**, 3250 (1975).
8. M. Raban, F. B. Jones, Jr., E. H. Carlson, E. Banucci, and N. A. LeBel, *J. Org. Chem.*, **35**, 1496 (1970).
9. N. A. LeBel, G. M. J. Slusarczuk, and L. A. Spurlock, *J. Amer. Chem. Soc.*, **84**, 4360 (1962).
10. N. A. LeBel, G. H. Greene, and P. R. Peterson, unpublished work.
11. N. A. LeBel, N. D. Ojha, J. R. Menke, and R. J. Newland, *J. Org. Chem.*, **37**, 2896 (1972).
12. W. Oppolzer and H. P. Weber, *Tetrahedron Lett.*, 1121 (1970).
13. W. Oppolzer and K. Keller, *Tetrahedron Lett.*, 4313 (1970).
14. N. A. LeBel, *Trans. N. Y. Acad. Sci.*, **27**, 858 (1965).
15. N. A. LeBel, T. A. Lajiness, and D. B. Ledlie, *J. Amer. Chem. Soc.*, **89**, 3076 (1967).

Appendix
Chemical Abstracts Nomenclature (Collective Index Number; Registry Numbers)

2,1-Benzisoxazoline, hexahydro-1,3,3,6-tetramethyl- (8); 2,1-Benzisoxazole, 1,3,3a,4,5,6,7,7a-octahydro-1,3,3,6-tetramethyl- (9); stereoisomer (6501-80-0); (6603-39-0)

6-Octenal, 3,7-dimethyl- (8,9); (106-23-0)

Hydroxylamine, N-methyl- (8); Methanamine, N-hydroxy- (9); (593-77-1)

Hydroxylamine, N-methyl-, hydrochloride (8); Methanamine, N-hydroxy-, hydrochloride (9); (4229-44-1)

Methane, nitro- (8,9); (75-52-5)

7-Octenal, 3,7-dimethyl- (8,9); (13827-93-5)

Lithium aluminum hydride: Aluminate (1-), tetrahydro-, lithium (8); Aluminate (1-), tetrahydro-, lithium (T-4)- (9); (16853-85-3)

NUCLEOPHILIC α-sec-AMINOALKYLATION: 2-(DIPHENYLHYDROXYMETHYL)PYRROLIDINE

(2-Pyrrolidinemethanol, α,α-diphenyl-)

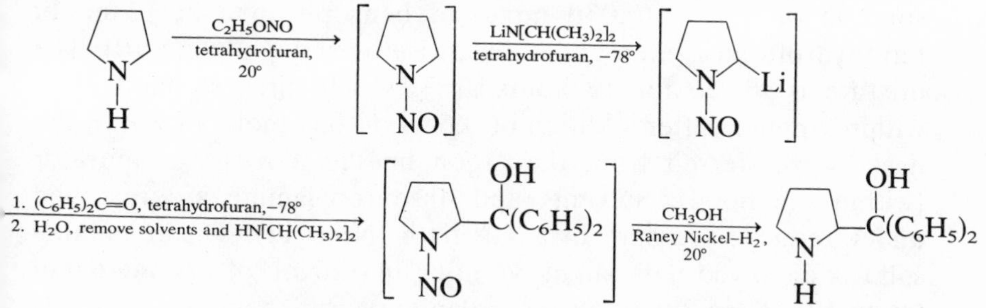

Submitted by D. Enders, R. Pieter, B. Renger, and D. Seebach[1]
Checked by C. Hutchins and M. F. Semmelhack

1. Procedure

Caution! Since N-nitrosopyrrolidine is a potent carcinogen and is produced as an intermediate, this entire "one-pot" procedure should be performed in a well-ventilated hood. Wearing of disposable polyethylene gloves is recommended.

A dry 250-ml., one-necked, round-bottomed flask is equipped with a magnetic stirrer and a three-way stopcock. The flask is charged with 4 g. (0.053 mole) of ethyl nitrite (Note 1), 4 g. of dry tetrahydrofuran (Note 2), and 2.35 g. (0.033 mole) of pyrrolidine (Note 3). The stopcock is closed (Note 4), and the mixture is stirred at room temperature for 2 days. Excess ethyl nitrite, tetrahydrofuran, and the ethanol formed are removed from the N-nitrosopyrrolidine (Note 5) by stirring at 25° under reduced pressure (10 mm., water aspirator, Note 6) for 2 hours. The stopcock is fitted with a rubber septum, the air in the system is replaced by dry argon (Notes 4 and 7), and 50 ml. of tetrahydrofuran is injected by syringe. A solution of lithium diisopropylamide is prepared in a separate dry 100-ml. flask by adding 21.1 ml. (0.034 mole) of a 1.61 M solution of butyllithium in hexane (Note 8) to a solution of 4.76 ml. (0.034 mole) of diisopropylamine (Note 9) in 25 ml. of tetrahydrofuran at −78° (dry ice–methanol bath) with

113

stirring under argon. The solution is warmed to 0° in 15 minutes and then added dropwise with a syringe within 4 minutes to the nitrosamine solution stirred at −78°. Stirring of the yellow to orange solution is continued at this temperature for 25 minutes. A solution of 5.46 g. (0.030 mole) of benzophenone in 12 ml. of tetrahydrofuran is then added dropwise by syringe (Note 10). The mixture is stirred for 12 hours at −78° and then warmed to 0° within 2 hours. After addition of 0.6 ml. (0.034 mole) of water, the flask is transferred from the argon line to a rotary evaporator (within the hood). Solvents and diisopropylamine are removed under reduced pressure in a 40° bath (Note 11). The remaining solid is dissolved with slight warming in 120 ml. of dry methanol (Note 12). Then 3.9 g. (66 equivalents) of Raney nickel (Note 13) is rinsed into the solution with 30 ml. of dry methanol. The reaction vessel is equipped again with the three-way stopcock, and the air in the flask is replaced by hydrogen (Note 7). The flask is filled five times with hydrogen from a balloon; during this operation vigorous stirring of the Raney nickel–methanol suspension is necessary. The flask is attached to a mercury bubbler to maintain 200 mm. of positive hydrogen pressure supplied from a cylinder, as shown in Figure 1. The reaction mixture is stirred for 3 hours at room temperature while a slow stream of hydrogen is passed through the system. The major part of the solution is decanted and filtered, and the remaining Raney-nickel suspension is extracted by

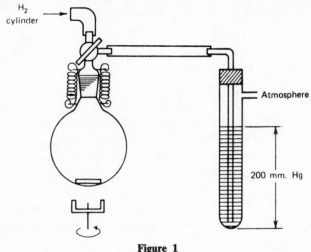

Figure 1

refluxing three times for 10 minutes each with 20 ml. of methanol (Note 14). The combined methanol solutions are concentrated under reduced pressure. The residue is dissolved in 150 ml. of ethyl ether and 100 ml. of water, the layers are separated (Note 15), and the aqueous layer is extracted three times with 50-ml. portions of ethyl ether. The combined extracts are dried over sodium carbonate and concentrated in a rotary evaporator to a total volume of 150 ml. Dry hydrogen chloride gas is bubbled into the solution with stirring until the mixture is acidic. The almost colorless precipitate of the hydrochloride is filtered, washed two times with 30-ml. portions of dry ethyl ether, and dried in a desiccator under reduced pressure for 3 hours. The product, m.p. 244–249° (dec.), weighs 5.99–6.11 g. (68–70% based on benzophenone). Analytically pure product is obtained by recrystallization from methanol–acetone, 5.06–5.20 g. (58–60%), m.p. 267–269° (dec.) (Note 16). The free base was obtained by treatment of the hydrochloride with 10% aqueous sodium hydroxide and extraction with ethyl ether to give colorless crystals, m.p. 82–83° (Note 17).

2. Notes

1. Ethyl nitrite is prepared as described in W. J. Semon and W. R. Damerell, *Org. Syn.*, Coll. Vol. **2**, 204 (1943), or purchased from Merck-Schuchardt and distilled before use, b.p. 17°. The volatile nitrite can be easily handled as a 50% tetrahydrofuran solution and stored in a refrigerator.

2. Technical-grade tetrahydrofuran, available from BASF-A G or Fisher Scientific Company, is dried by distillation, first from potassium hydroxide and then from lithium aluminum hydride. This tetrahydrofuran is used for all operations in this procedure. For a warning note regarding the purification of tetrahydrofuran, see *Org. Syn.*, Coll. Vol. **5**, 976 (1973).

3. Pyrrolidine, b.p. 87–88°, obtained from Aldrich Chemical Company, Inc. or BASF-A G is distilled from potassium hydroxide before use.

4. The three-way stopcock with standard-tapered joint must be securely fastened to the neck of the flask with wire, rubber band, or springs [see Figure 1 and D. Seebach and A. K. Beck, *Org. Syn.*, **51**, 39, 76 (1971)].

5. Nitrosamines are strong carcinogens[2,3]; N-nitrosopyrrolidine causes liver tumors in rats.[2,4] Although the one-pot procedure described here prevents contact with the nitrosamine, utmost care must be used to avoid contact with the reaction mixture during all manipulations.

6. At the beginning of the evacuation the pressure should be lowered slowly to prevent bumping.

7. This is done by alternately evacuating and filling with dry argon three times; during the reaction a pressure of about 50 mm. above atmospheric is maintained using a mercury bubbler.

8. Purchased from Metallgesellschaft, Frankfurt, or Alfa-Products, Division of the Ventron Corporation. The content of the solution was determined prior to use by acidimetric titration.

9. The diisopropylamine, b.p. 83–84°, available from Fluka A G, BASF-A G, or Aldrich Chemical Company, Inc., is purified by refluxing over potassium hydroxide and subsequent distillation. It is stored over calcium hydride.

10. Benzophenone, m.p. 47–49°, was purchased from Riedel-de-Haen-A G or from Fisher Scientific Company. The reaction mixture turns green and then blue during the addition, because of the formation of ketyl radicals.

11. The checkers found it more convenient to remove the volatile material at this stage by warming the stirred mixture at 40° using a water aspirator vacuum (10 mm.). About 3 hours were required.

12. Methanol was dried by heating at reflux for 3 hours over magnesium and then distilling.

13. The Raney nickel reagent is prepared by addition of 9.5 g. of sodium hydroxide pellets over 8–10 minutes to a stirred suspension of 7.8 g of nickel–aluminum alloy (50% Ni, 50% Al powder, purchased from Merck-Schuchardt) in 120 ml. of distilled water contained in a 250-ml. beaker. Fifteen minutes after the addition is completed, the beaker is immersed into a 70° water bath for 20 minutes. The water is decanted, and the catalyst is washed sequentially with two 20-ml. portions of distilled water and two 20-ml. portions of methanol.

14. *Caution! The dry Raney nickel catalyst is pyrophoric. The residues can be destroyed by allowing them to ignite and burn on filter paper in a safe place.*

15. Upon dissolving the residue in 200 ml. of ethyl ether and 100 ml. of water, the checkers obtained an emulsion that cleared slowly on standing for 2–3 hours.

16. The literature reports m.p. >240°,[5] >250°,[6] and 262–263°.[7] The yield given includes a small second crop obtained by recrystallization of the filtrate residue. The reported yield and m.p. data were obtained by the checkers. The submitters report 6.50–6.95 g. (75–80%, m.p. 260–265°) before recrystallization and 5.20–5.62 g. (60–65%, m.p. 267–269°) for analytically pure product.

17. The m.p. is reported to be 81–82°[6] and 83°.[5] The spectral properties are: infrared spectrum (KI) cm^{-1}: strong absorptions at 3360, 3080, 3060, 3020, 2980–2800, 1595, 1490, 1450, 1400, 1190, 1100, 1060, 1030, 990, 900, 750, 700, 660, and 635; proton magnetic resonance (chloroform-d) δ (multiplicity, number of protons): 7.65–7.00 (multiplet, 10), 4.18 (multiplet, 1), 2.95 (multiplet, 2), with an overlapping broad peak (for OH and NH), and 1.60 (multiplet, 4). The compound has psychostimulating activity.[7]

3. Discussion

The procedure described here is an example of overall conversions by the "nitrosamine method" for the electrophilic substitution of **1** to **3** via the intermediate anion **2**, as outlined in detail in a recent review article.[8]

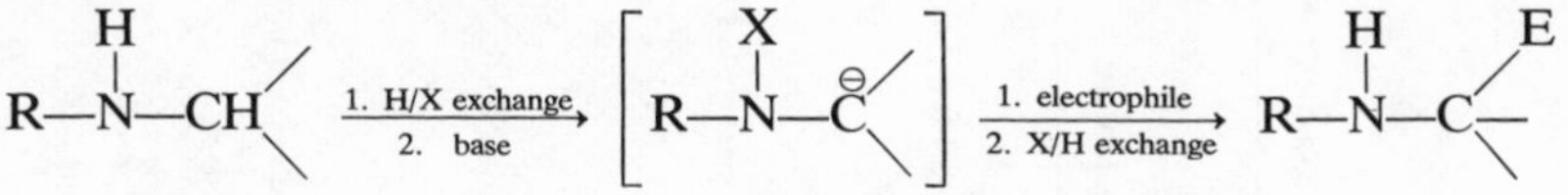

This is currently the only method that allows reversible enhancement of the acidity of α-nitrogen CH-protons in a large variety of secondary amines using the nitroso group on the nitrogen (X = NO, see examples in Table I). The nucleophile **2** is synthetically equivalent to an α-aminocarbanion **4** while the inherent reactivity at carbons next to amino nitrogens is electrophilic (see the immonium ion **5**).[8,9] Lithiated nitrosamines are also useful because their

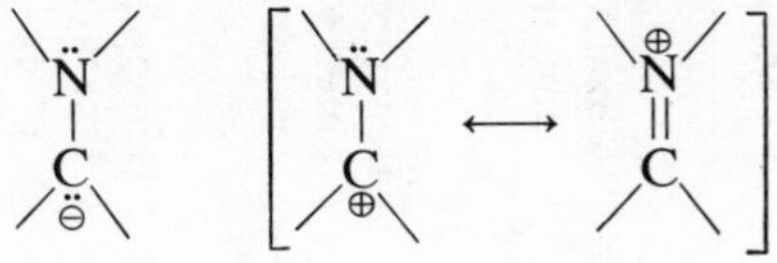

TABLE I

α-Substituted Secondary Amines via Electrophilic Substitution[8]

Starting Materials	Nitrosamine	Metallation Time (minutes)	Product[a]	Yield (%)
Benzaldehyde, dimethylamine	CH_3—N—CH_3 $\mid$ NO	10	CH_3—N—CH_2—CH—C_6H_5 (OH on CH; H on N)	80
Benzylbromide, methylisopropylamine	$(CH_3)_2CH$—N—CH_3 $\mid$ NO	10	$(CH_3)_2CH$—N—CH_2—CH_2—C_6H_5 (H on N)	80
Benzophenone, methylisopropylamine	$(CH_3)_2CH$—N—CH_3 $\mid$ NO	10	$(CH_3)_2CH$—N—CH_2—$C(C_6H_5)_2$ (OH; H on N)	75
Piperonal, methyl tert-butylamine	$(CH_3)_3C$—N—CH_3 $\mid$ NO	10	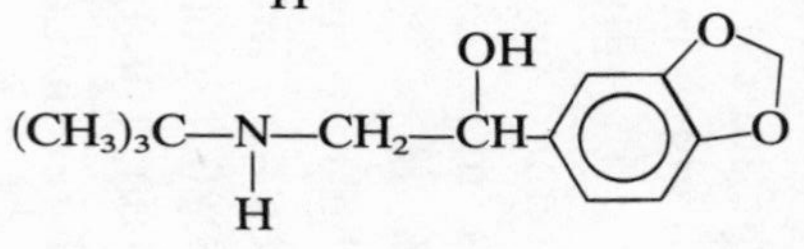 $(CH_3)_3C$—N—CH_2—CH (OH; H on N) methylenedioxyphenyl	80
Carbon dioxide, methyl tert-butylamine	$(CH_3)_3C$—N—CH_3 $\mid$ NO	10	$(CH_3)_3C$—N—CH_2—COOH (H on N)	75[b]

Benzaldehyde, diethylamine	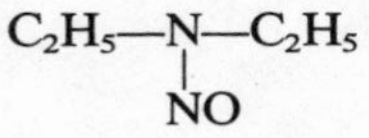 C_2H_5—N—C_2H_5 NO	10	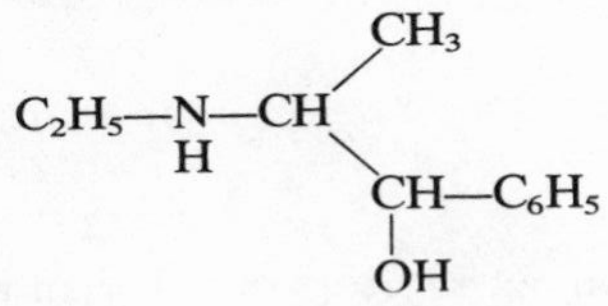 C_2H_5—N—CH H CH₃ CH—C_6H_5 OH	75[b]
Benzaldehyde, dihexylamine	 C_6H_{13}—N—C_6H_{13} NO	10	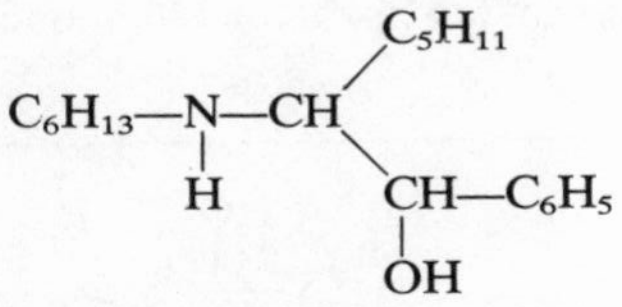 C_6H_{13}—N—CH H C_5H_{11} CH—C_6H_5 OH	80[b]
Benzophenone, azetidine	 (azetidine)N—NO	7	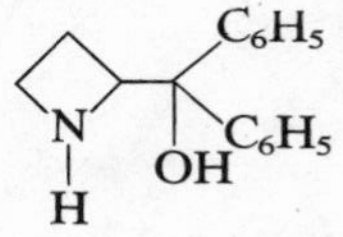 (azetidine)N—H with C(C_6H_5)(C_6H_5)OH	75[b]
Benzophenone, piperidine	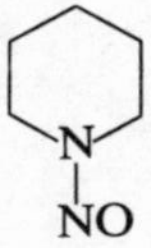 (piperidine)N—NO	180	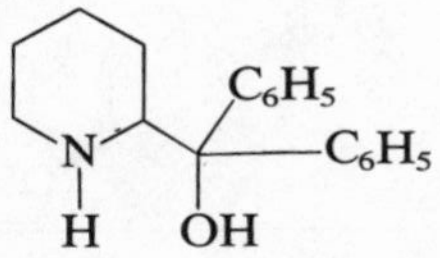 (piperidine)N—H with C(C_6H_5)(C_6H_5)OH	50
Cyclohexanone, piperidine	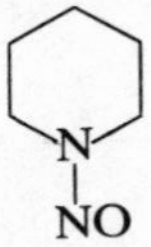 (piperidine)N—NO	180	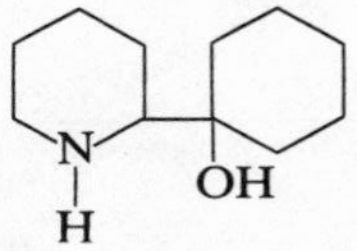 (piperidine)N—H with C(cyclohexyl)OH	55

119

continued overleaf

TABLE I (cont.)

Starting Materials	Nitrosamine	Metallation Time (minutes)	Product[a]	Yield (%)
Iodopropane, 3-hydroxy-piperidine	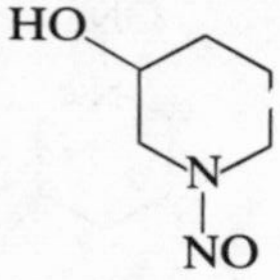	240	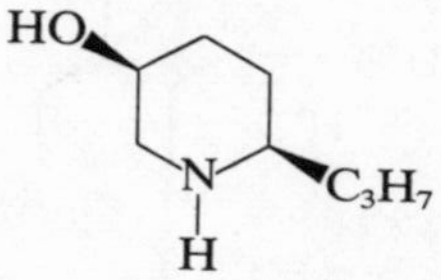	40
Benzophenone, perhydroazepine	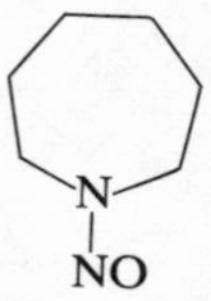	20	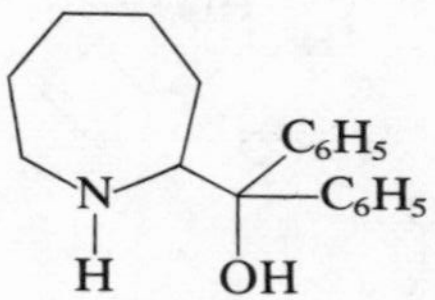	80[b]

[a] Isolated and characterized as hydrochlorides.

[b] Overall yield of stepwise procedure; the denitrosation was performed by bubbling gaseous hydrogen chloride into a benzene solution of the nitrosamine. This cleavage procedure of nitrosamines is usually not as clean as the one with Raney nickel.

formation and reactions with electrophiles occur with high yields as well as with a high degree of regio- and stereoselectivity (see Table I). For further information see the review article[8] and other recent publications.[10] This method has the disadvantage of carcinogenic nitrosamine intermediates, but this one-pot procedure reduces the danger of contact.

1. Institüt für Organische Chemie der Justus Liebig-Universität, 6300 Giessen, Germany. Dr. Seebach's present address: Eidg. Techische Hochschule, Laboratorium für Organische Chemie, Universitätsstrabe 16, CH-8092 Zürich, Switzerland.
2. H. Druckrey, R. Preussmann, S. Ivankovic, and D. Schmaehl, Z. Krebsforsch, 69, 103 (1967).
3. P. N. Magee and J. M. Barnes, Advan. Cancer Res., 10, 163 (1967).
4. M. Greenblatt and W. Lijinsky, J. Nat. Cancer Inst., 48, 1687 (1972).
5. J. Kapfhammer and A. Matthes, Hoppe-Seylers Z. Physiol. Chem., 223, 43 (1933).
6. S. O. Winthrop and L. G. Humber, J. Org. Chem., 26, 2834 (1961).
7. A. M. Likhosherstov, K. S. Raevskii, A. S. Lebedeva, A. M. Kritsyn, and A. P. Skoldinov, Khim. Farm. Zh., 1, 30 (1967) [C.A., 67, 90,642w (1967)].
8. Review: D. Seebach and D. Enders, Angew. Chem., 87, 1 (1975); Angew. Chem. Int. Ed. Engl., 14, 15 (1975). Preliminary communications: D. Seebach and D. Enders, Angew. Chem., 84, 350, 1186, 1187 (1972); Angew. Chem. Int. Ed. Engl., 11, 301, 1101, 1102 (1972); D. Seebach, D. Enders, B. Renger, and W. Bruegel, Angew. Chem., 85, 504 (1973); Angew. Chem. Int. Ed. Engl., 12, 495 (1973). Full papers with experimental detail: D. Seebach and D. Enders, J. Med. Chem., 17, 1225 (1974); Chem. Ber., 108, 1293 (1975); D. Seebach, D. Enders, and B. Renger, Chem. Ber., 110, 1852 (1977); B. Renger, H.-O. Kalinowski, and D. Seebach, Chem. Ber., 110, 1866 (1977); D. Seebach, D. Enders, R. Dach, and R. Pieter, Chem. Ber., 110, 1879 (1977); B. Renger and D. Seebach, Chem. Ber., 110, 2334 (1977).
9. D. Seebach and M. Kolb, Chem. Ind. London, 687 (1974), 910 (1975).
10. R. R. Fraser and T. B. Grindley, Can. J. Chem., 53, 2465 (1975); R. R. Fraser, T. B. Grindley, and S. Passannanti, Can. J. Chem., 53, 2473 (1975); D. H. R. Barton, R. D. Bracho, A. A. L. Gunatilaka, and D. A. Widdowson, J. Chem. Soc., Perkin Trans. I, 579 (1975); J. E. Baldwin, S. E. Branz, R. F. Gomez, P. L. Kraft, A. J. Sinskey, and S. R. Tannenbaum, Tetrahedron Lett., 333 (1976). R. R. Fraser and S. Passananti, Synthesis, 540 (1976).

Appendix
Chemical Abstracts Nomenclature (Collective Index Number; Registry Numbers)

2-(Diphenylhydroxymethyl)pyrrolidine: 2-Pyrrolidinemethanol, α,α-diphenyl-, (+)- (8,9); (22348-32-9)

Pyrrolidine (8,9); (123-75-1)

N-Nitrosopyrrolidine: Pyrrolidine, 1-nitroso- (8,9); (930-55-2)

Ethyl nitrite: Nitrous acid, ethyl ester (8,9); (109-95-5)

Lithium diisopropylamide: Diisopropylamine, lithium salt (8); 2-Propanamine, N-(1-methylethyl)-, lithium salt (9); (4111-54-0)

Lithium, butyl- (8,9); (109-72-8)

Diisopropylamine (8); 2-Propanamine, N-(1-methylethyl)- (9); (108-18-9)

Benzophenone (8); Methanone, diphenyl- (9); (119-61-9)

Lithium aluminum hydride: Aluminate (1-), tetrahydro-, lithium (8); Aluminate (1-), tetrahydro-, lithium, (T-4)- (9); (16853-85-3)

OXIDATION OF ALCOHOLS BY METHYL SULFIDE-N-CHLOROSUCCINIMIDE-TRIETHYLAMINE: 4-*tert*-BUTYLCYCLOHEXANONE

[Cyclohexanone, 4-(1,1-dimethylethyl)-]

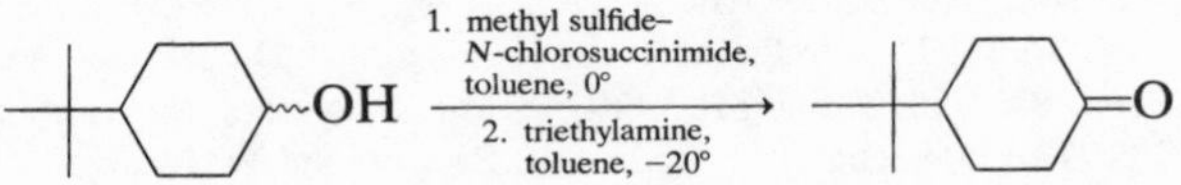

Submitted by E. J. COREY,[1] C. U. KIM,[2] and P. F. MISCO[2]
Checked by S. YAMAMOTO and W. NAGATA

1. Procedure

Caution! See benzene warning, p. 168.

A 1-l., three-necked, round-bottomed flask is equipped with a mechanical stirrer (Note 1), a thermometer, a dropping funnel, and an argon-inlet tube. The flask is charged with 8.0 g. (0.060 mole) of N-chlorosuccinimide (Notes 2 and 3) and 200 ml. of toluene (Note 4). While a continuous positive argon pressure is maintained, the solution is stirred and cooled to 0°, and 6.0 ml. (0.10 mole) of methyl sulfide (Note 2) is added. A white precipitate appears immediately after addition of the sulfide. The mixture is cooled to −25° using a carbon tetrachloride–dry ice bath (Note 5), and a solution of 6.24 g. (0.04 mole) of 4-*tert*-butylcyclohexanol (mixture of *cis* and *trans*) (Note 2) in 40 ml. of toluene is added dropwise over 5 minutes (Note 6). The stirring is continued for 2 hours at −25° (Note 5), and then a solution of 6.0 g. (0.06 mole) of triethylamine (Note 2) in 10 ml. of toluene is added dropwise over 3 minutes (Note 7). The cooling bath is removed; after 5 minutes 400 ml. of ethyl ether is added (Note 8). The organic layer is washed with 100 ml. of 1% aqueous hydrochloric acid and then twice with two 100-ml. portions of water (Note 9) and dried over anhydrous magnesium sulfate. The solvents are evaporated under reduced pressure. The residue is transferred to a 50-ml., round-bottomed flask and distilled bulb to bulb at 120° (25 mm.) to yield 5.54–5.72 g. (90–93%) of 4-*tert*-butylcyclohexanone, m.p. 41–45° (Note 10).

2. Notes

1. Efficient magnetic stirring could be used as well.

2. The submitters used N-chlorosuccinimide, 4-*tert*-butyl-cyclohexanol, methyl sulfide, and triethylamine, available from Aldrich Chemical Company, Inc., without further purification.

The checkers used *cis*- and *trans*-4-*tert*-butylcyclohexanol (36:64) obtained from E. Merck A G, Darmstadt, methyl sulfide obtained from Tokyo Kasei Kogyo Co., Ltd., and triethylamine obtained from Wako Pure Chemical Industries, Ltd., without further purification.

3. The checkers observed that, when N-chlorosuccinimide of 93–95% purity, m.p. 128–132° (Checked by iodometry), obtained from E. Merck A G, Darmstadt, was used without purification, the oxidation was incomplete, resulting in 93–94% yields of a product containing a 12–15% amount of the starting alcohol (a mixture of the *cis* and *trans*). The use of 98% pure N-chlorosuccinimide of m.p. 150–151° (recrystallized from benzene) resulted again in recovery of the alcohol in a considerable amount, mainly because of low solubility of the pure reagent in toluene. Therefore, the checkers slightly modified the earlier part of the procedure as follows: the 98% pure N-chlorosuccinimide (8.0 g., 0.06 mole) m.p. 150–151°, is dissolved in 400 ml. of toluene (twice as much as the volume used by the submitters) at 40°, and the solution is cooled to room temperature, stirred under nitrogen atmosphere, and then cooled to 0–5°. Methyl sulfide (6 ml., 0.10 mole) is added at this temperature and, after addition, the reaction mixture is stirred for an additional 20 minutes at 0–3°, during which time white precipitate appeared.

4. Reagent-grade toluene employed by the submitters was obtained from Mallinckrodt Chemical Works. The checkers used reagent-grade toluene purchased from Wako Pure Chemical Industries, Ltd., and dried over molecular sieves (4 Å) before use. Regarding the volume of the toluene used by the checkers, see Note 3).

5. The checkers observed that the internal temperature was −20° with this cooling mixture.

6. The checkers observed that the reaction was slightly exothermic as judged from an internal temperature rise of *ca.* 5°.

7. The checkers needed a little longer time (3–5 minutes) for this operation to maintain the internal temperature at −15 to −20°.

8. The checkers successively added 100 ml. 1% aqueous hydrochloric acid to the reaction mixture.

9. Vigorous shaking is necessary for complete removal of succinimide. The triethylamine hydrochloride is also removed in this step. The methyl sulfide codistils with the ethyl ether.

10. An authentic sample from Aldrich Chemical Company, Inc., had m.p. 45–50°. The product is analyzed by gas chromatography at 80° on F & M Research chromatograph model 1810 with 3% OV-17 column, which indicates the contamination of the product by 4-*tert*-butylcyclohexanol (<2%) and 4-*tert*-butylcyclohexylmethyl methylthiomethyl ether (<2%).[3] The submitters reported a yield of 6.0 g. (96%) of purity greater than 96%.

The checkers recrystallized the product from petroleum ether (dissolved at room temperature and cooled to −20°) and obtained a pure product of m.p. 45–46° (88.5% recovery). The pure product has the following spectral data: infrared (chloroform) cm^{-1}: 1712 (C=O); proton magnetic resonance (chloroform-d) δ (multiplicity, number of protons, assignment): 0.93 [singlet, 9, C(CH$_3$)$_3$], 1.3–2.2 (multiplet, 5, CH_2—CH—CH_2), and 2.2–2.50 (multiplet, 4, CH$_2$—CO—CH$_2$).

3. Discussion

The procedure described here is general for the oxidation of primary and secondary alcohols to carbonyl compounds,[3] but not for allylic and dibenzylic alcohols, which give halides in high yields.[4] The yields of carbonyl compounds are usually high, and the

TABLE I
OXIDATION OF PRIMARY AND SECONDARY ALCOHOLS[3]

Alcohol	Product	Yield (%)
$C_6H_5CH_2OH$	C_6H_5CHO	90
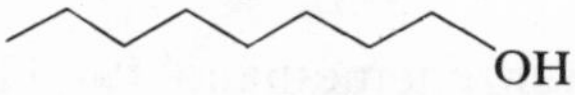 (n-heptyl alcohol, structure with OH)	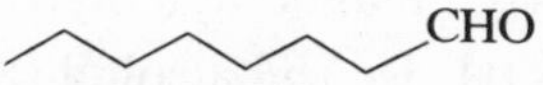 (aldehyde, structure with CHO)	96
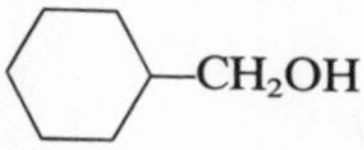 (cyclohexyl–CH_2OH)	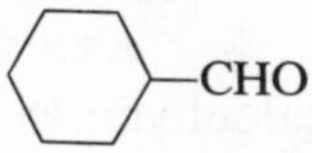 (cyclohexyl–CHO)	94
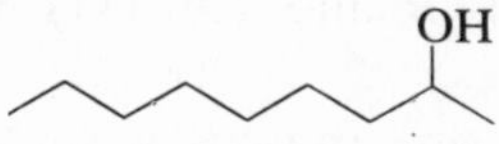 (secondary alcohol, OH)	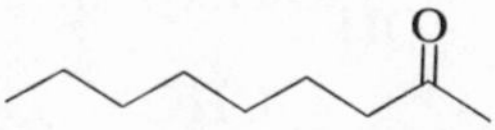 (ketone, O)	91
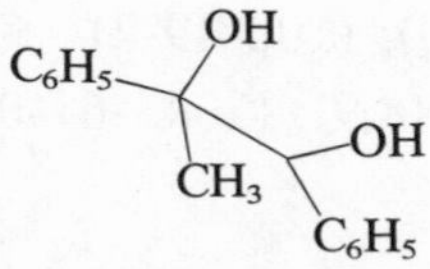 (C_6H_5, OH, OH, CH_3, C_6H_5)	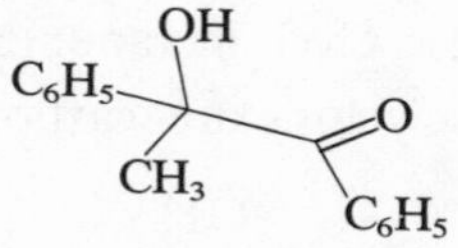 (C_6H_5, OH, CH_3, O, C_6H_5)	86[5]

formation of methyl thiomethyl ether can be minimized in nonpolar media. Some examples are listed in Table I. The quantitative conversion of catechols to o-quinones using this oxidation procedure is reported.[6] For oxidation of allylic and dibenzylic alcohols to the corresponding carbonyl compounds, a dimethyl sulfoxide–chlorine reagent[7] is suitable.

1. Department of Chemistry, Harvard University, Cambridge, Massachusetts 02138.
2. Research Division, Bristol Laboratories, Syracuse, New York 13201.
3. E. J. Corey and C. U. Kim, *J. Amer. Chem. Soc.*, **94**, 7586 (1972); E. J. Corey and C. U. Kim, *J. Org. Chem.*, **38**, 1233 (1973).
4. E. J. Corey, C. U. Kim, and M. Takeda, *Tetrahedron Lett.*, 4339 (1972).
5. E. J. Corey and C. U. Kim, *Tetrahedron Lett.*, 287 (1974).
6. J. P. Marino and A. Schwatz, *J. Chem. Soc., Chem. Commun.*, 812 (1974).
7. E. J. Corey and C. U. Kim, *Tetrahedron Lett.*, 919 (1973).

Appendix
Chemical Abstracts Nomenclature (Collective Index Number; Registry Numbers)

N-Chlorosuccinimide: Succinimide, N-chloro- (8); 2,5-Pyrrolidinedione, 1-chloro- (9); (128-09-6)

Toluene (8); Benzene, methyl- (9); (108-88-3)

Methyl sulfide (8); Methane, thiobis- (9); (75-18-3)

Cyclohexanol, 4-*tert*-butyl- (8); Cyclohexanol, 4-(1,1-dimethylethyl)- (9); (98-52-2)

Triethylamine (8); Ethanamine, *N,N*-diethyl- (9); (121-44-8)

4-*tert*-Butylcyclohexanone: Cyclohexanone, 4-*tert*-butyl- (8); Cyclohexanone, 4-(1,1-dimethylethyl)- (9); (98-53-3)

C_6H_5—CH_2—OH: Benzyl alcohol (8); Benzenemethanol (9); (100-51-6)

C_6H_5—CHO: Benzaldehyde (8,9); (100-52-7)

: Octyl alcohol (8); 1-Octanol (9); (111-87-5)

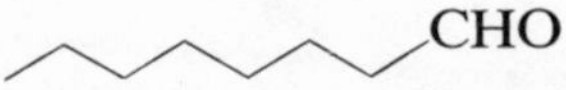

: Octanal (8, 9); (124-13-0)

C_6H_{11}—CH_2OH: Cyclohexanemethanol (8,9); (100-49-2)

C_6H_{11}—CHO: Cyclohexanecarboxaldehyde (8,9); (2043-61-0)

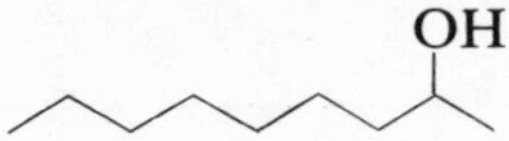

: 2-Nonanol (8, 9); (628-99-9)

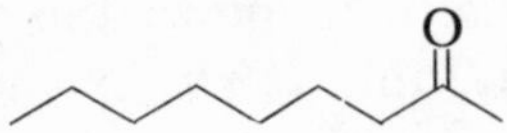

: 2-Nonanone (8, 9); (821-55-6)

C_6H_5—C—CH—C_6H_5: 1,2-Propanediol, 1,2-diphenyl- (8,9); (41728-16-9)

C_6H_5—C—C—C_6H_5: Benzoin, α-methyl- (8); 1-Propanone, 2-hydroxy-1,2-diphenyl- (9); (5623-26-7)

Triethylamine, hydrochloride (8); Ethanamine, *N,N*-diethyl-, hydrochloride (9); (554-68-7)

Succinimide (8); 2,5-Pyrrolidinedione (9); (123-56-8)

tert-Butylcyclohexylmethyl methylthiomethyl ether: Cyclohexane, 1-*tert*-butyl-4-(methylthiomethoxymethyl)- (8); Cyclohexane, 1-(1,1-dimethylethyl)-4-(methylthiomethoxymethyl)- (9); (—)

PHOSPHINE–NICKEL COMPLEX CATALYZED CROSS-COUPLING OF GRIGNARD REAGENTS WITH ARYL AND ALKENYL HALIDES: 1,2-DIBUTYLBENZENE

$$CH_3(CH_2)_3Br + Mg \xrightarrow{\text{ethyl ether}} CH_3(CH_2)_3MgBr$$

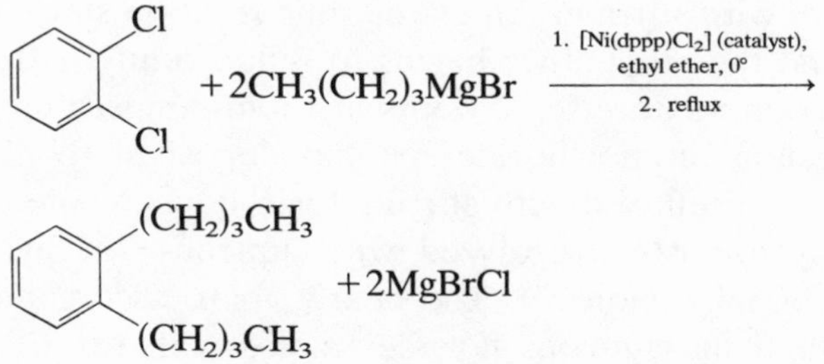

$$dppp = (C_6H_5)_2P(CH_2)_3P(C_6H_5)_2$$

Submitted by Makoto Kumada, Kohei Tamao, and Koji Sumitani[1]
Checked by Teresa Y. L. Chan and S. Masamune

1. Procedure

A 500-ml., three-necked flask is equipped with a mechanical stirrer, a pressure-equalizing dropping funnel, and a reflux condenser attached to a nitrogen gas inlet. The flask is charged with 12.2 g. (0.50 g.-atom) of magnesium turnings. The magnesium is dried under a rapid stream of nitrogen with a heat gun. After the flask has cooled to room temperature, the rate of nitrogen flow is reduced, and 200 ml. of anhydrous ethyl ether (Note 1) and approximately 5 ml. of a solution of 68.5 g. (0.50 mole) of 1-bromobutane (Note 2) in 50 ml. of anhydrous ethyl ether are added. The mixture is stirred at room temperature, and within a few minutes an exothermic reaction occurs. The flask is immersed in an ice bath, and the remaining ether solution is added dropwise over *ca.* 1 hour. After addition is complete, the mixture is refluxed with stirring for 30 minutes and then cooled to room temperature.

A 1-l., three-necked flask, equipped in the same manner, is charged with *ca.* 0.25 g. (*ca.* 0.5 mmole) of dichloro[1,3-bis(diphenylphosphino)propane]nickel(II) (Note 3), 29.5 g. (0.20

mole) of 1,2-dichlorobenzene (Note 4), and 150 ml. of anhydrous ethyl ether (Note 5). The Grignard reagent prepared
above is transferred to the dropping funnel and added over
10 minutes, with stirring, to the mixture cooled in an ice bath. The
nickel complex reacts immediately with the Grignard reagent, and
the resulting clear-tan reaction mixture is allowed to warm to room
temperature with stirring. An exothermic reaction starts within 30
minutes, and the ethyl ether begins to reflux gently. After stirring
of the reaction mixture for 2 hours at room temperature, most of
the magnesium bromochloride salt has deposited (Note 6). The
mixture is then refluxed with stirring for 6 hours, cooled in an ice
bath, and cautiously hydrolyzed with aqueous $2N$ hydrochloric
acid (*ca.* 250 ml.) (Note 7). The nearly colorless organic layer is
separated and the aqueous layer extracted with two 70-ml. portions of ethyl ether. The combined organic layer and extracts are
washed successively with water, aqueous saturated sodium bicarbonate, and again with water, then dried over anhydrous calcium
chloride and filtered. After evaporation of the solvent the residue
is distilled under reduced pressure through a 25-cm. column
packed with glass helices to give a forerun (*ca.* 4 g.) of 1-butyl-2-
chlorobenzene, b.p. 52–54° (3.5 mm.), n^{20}D 1.5110, followed by
30.0–31.5 g. (79–83%) of 1,2-dibutylbenzene, b.p. 76–81°
(3.5 mm.), n^{20}D 1.4920, as a colorless liquid (Note 8).

2. Notes

1. Ethyl ether is dried over sodium wire and freshly distilled
before use.

2. Commercial 1-bromobutane is dried over anhydrous calcium
chloride and distilled before use.

3. Dichloro[1,3-bis(diphenylphosphino)propane]nickel(II) can be
easily prepared in an open reaction vessel.[2] To a hot solution of
9.5 g. (0.040 mole) of nickel(II) chloride hexahydrate in 175 ml. of
a 5:2 (*v*/*v*) mixture of 2-propanol and methanol, is added, with
stirring, a hot solution of 14.5 g. (0.035 mole) of 1,3-
bis(diphenylphosphino)propane in 200 ml. of 2-propanol. A red-
dish brown precipitate deposits immediately. The mixture is heated
for 30 minutes and allowed to cool to room temperature. Filtration, washing with methanol, and drying under reduced pressure

provide the red complex in almost quantitative yield. 1,3-Bis(diphenylphosphino)propane may be purchased from Strem Chemicals Inc., and nickel(II) chloride hexahydrate from Fisher Scientific Company.

4. Commercial G.R.-grade 1,2-dichlorobenzene can be used without further purification. At least a 20% excess of the Grignard reagent should be used to compensate for some loss through undesirable side reactions (see Discussion section).

5. The nickel complex is insoluble in the mixture.

6. If stirring becomes difficult, approximately 100 ml. of anhydrous ethyl ether may be added.

7. This hydrolysis is exothermic, and the acid should be added slowly so as to maintain gentle refluxing of the ethyl ether.

8. Gas chromatographic analysis on a 3.5-m. column packed with Silicone DC 550 and operated at 200° showed that the product was at least 99.5% pure. The product has the following spectral properties: infrared (neat): $750 \, cm^{-1}$ (1,2-disubstituted benzene); proton magnetic resonance (carbon tetrachloride) δ (multiplicity; number of protons, assignment): 7.00 (singlet, 4, aromatic), 2.60 (triplet, 4, benzylic CH_2), 1.7–0.7 (multiplet, 14, $CH_2CH_2CH_3$).

3. Discussion

1,2-Dibutylbenzene has been prepared from cyclohexanone by tedious, multistep procedures.[3,4] The present one-step method is based on the selective cross-coupling of a Grignard reagent with an organic halide in the presence of a phosphine–nickel catalyst.[5]

The phosphine–nickel complex catalyzes the cross-coupling of alkyl, alkenyl, aryl, and heteroaryl Grignard reagents with aryl, heteroaryl, and alkenyl halides. The method thus has wide application. Alkyl halides also exhibit considerable reactivity, but give a complex mixture of products. Some representative examples are listed in Table I. A labile diorganonickel complex involving the two organic groups originating from the Grignard reagent and from the organic halide, respectively, has been proposed as a reaction intermediate. The fact that simple alkyl Grignard reagents with hydrogen-bearing β-carbon atoms react with equal efficiency is one of the most remarkable features of the present method,

TABLE I

CROSS-COUPLING OF GRIGNARD REAGENTS WITH VARIOUS ARYL AND ALKENYL HALIDES IN THE PRESENCE OF [Ni(dppp)Cl$_2$] AS A CATALYST[a]

Entry	RMgX	Halides	Product (% yield)[b]
1	CH$_3$MgBr	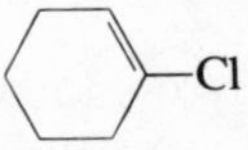 ⬡—Cl	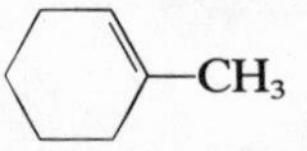 ⬡—CH$_3$ (98)
2	CH$_3$(CH$_2$)$_3$MgBr	C$_6$H$_5$Cl	C$_6$H$_5$C$_4$H$_9$ (96)
3	CH$_3$(CH$_2$)$_3$MgBr	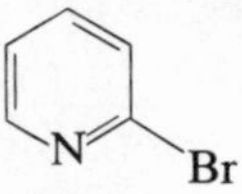 (pyridine)-Br	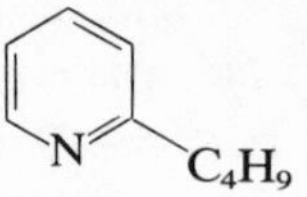 (pyridine)-C$_4$H$_9$ (71)[c]
4	(CH$_3$)$_3$SiCH$_2$MgCl	1,3-Cl$_2$C$_6$H$_4$	1,3-[(CH$_3$)$_3$SiCH$_2$]$_2$C$_6$H$_4$ (83)[c]
5	(CH$_3$)$_2$CHMgCl	C$_6$H$_5$Cl	(CH$_3$)$_2$CH- and CH$_3$(CH$_2$)$_2$C$_6$H$_5$ (96:4) (89) (10:90) (84)[d]
6	C$_6$H$_5$CH(CH$_3$)MgCl	CH$_2$=CHCl[e,f]	(R)-(−)-C$_6$H$_5$CH(CH$_3$)CH=CH$_2$[g] (81)
7	BrMg(CH$_2$)$_{10}$MgBr	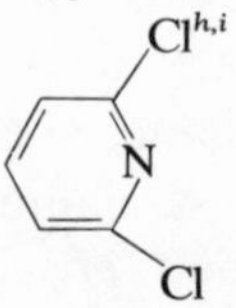 2,6-Cl$_2$-pyridine[h,i]	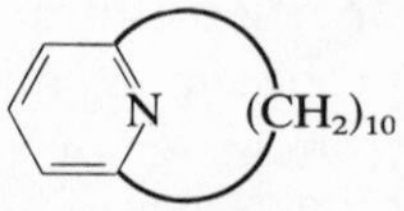 pyridine-(CH$_2$)$_{10}$ bridge (33)
8	C$_6$H$_5$MgBr	(Z)-C$_6$H$_5$CH=CHBr	(Z)-C$_6$H$_5$CH=CHC$_6$H$_5$ (100)
9	C$_6$H$_5$MgBr	ClCH=CHCl[d,j]	(Z)-C$_6$H$_5$CH=CHC$_6$H$_5$[k] (95–100)
10	4-ClC$_6$H$_4$MgBr	CH$_2$=CHCl[l]	4-ClC$_6$H$_4$CH=CH$_2$ (79)[c]
11	2,4,6-(CH$_3$)$_3$C$_6$H$_2$MgBr	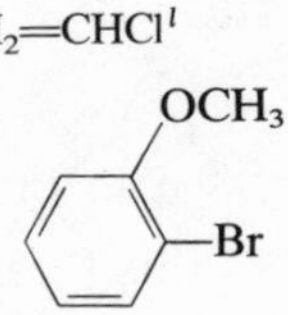 2-OCH$_3$-phenyl-Br	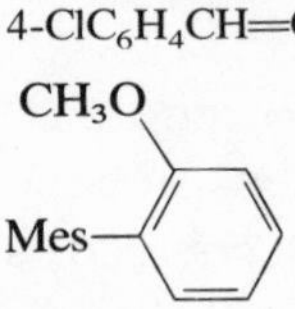 2-CH$_3$O-phenyl-Mes (74)
12	CH$_2$=C(CH$_3$)MgBr	α-C$_{10}$H$_7$Br[d,h]	α-C$_{10}$H$_7$C(CH$_3$)=CH$_2$ (78)
13	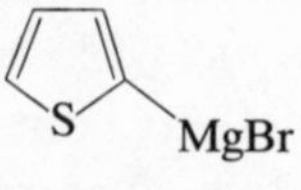 (thiophene)-MgBr	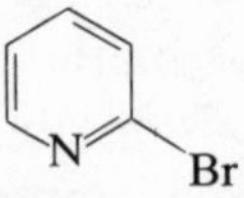 2-pyridine-Br	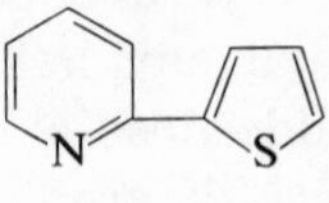 (pyridine)-(thiophene) (78)

[a] The reaction was carried out on a 0.01–0.02 mole scale in refluxing ethyl ether for 3–20 hours, unless otherwise noted.
[b] Determined by gas chromatography, unless otherwise noted.
[c] Isolated Yield.
[d] The catalyst is [Ni(dmpe)Cl$_2$], dmpe = (CH$_3$)$_2$PCH$_2$CH$_2$P(CH$_3$)$_2$.

130

considering the great tendency with which transition metal alkyls undergo a β-elimination reaction to form an alkene and a metal hydride.[6] Although the β-elimination may be responsible for the side products formed during the coupling, an appropriate choice of phosphine ligand for the nickel catalyst minimizes this side reaction. The catalytic activity of the phosphine–nickel complex depends not only on the nature of the phosphine ligand but also on the combination of the Grignard reagent and organic halide. Of several catalysts, including [Ni(dppp)Cl$_2$], [Ni(dppe)Cl$_2$], [Ni(dmpe)Cl$_2$], and [Ni{P(C$_6$H$_5$)$_3$}$_2$Cl$_2$], where dppe $= (C_6H_5)_2$-PCH$_2$CH$_2$P(C$_6$H$_5$)$_2$ and dmpe $= (CH_3)_2$PCH$_2$CH$_2$P(CH$_3$)$_2$, the first has been found most effective in almost all cases, except for alkenyl Grignard reagents. The halide may be chloride, bromide, or iodide, although chlorides usually give the most satisfactory results. Even fluorides react with comparable facility in some cases.[7] Vinyl chloride is one of the most reactive halocompounds, and the coupling (Table I, entry 10) can be conveniently conducted in an open system, similar to that described here, by the gradual addition of the Grignard reagent at 0° to a mixture of vinyl chloride, [Ni(dppp)Cl$_2$], and ethyl ether, followed by stirring at room temperature for 2 hours.

Several other features deserve comment. The coupling of secondary alkyl Grignard reagents is accompanied by alkyl group isomerization from secondary to primary, the extent of which is strongly dependent on the electronic nature of the phosphine ligand in the catalyst (entry 5).[7] Asymmetric cross-coupling can be achieved by using optically active phosphine–nickel complexes as catalysts (entry 6).[8,9] Cyclocoupling of di-Grignard reagents with dihalides offers a new, one-step route to cyclophanes (entry 7).[10]

---

[e] The catalyst is [Ni{(−)-diop}Cl$_2$], (−)-diop = 2,3-O-isopropylidene-2,3-dihydroxy-1,4-bis(diphenylphosphino)butane (see reference 12).
[f] At 0° for 2 hours.
[g] Configuration of the predominant enantiomer; 13.0% enantiomeric excess.
[h] Solvent is tetrahydrofuran.
[i] At 40° for 9 hours.
[j] At room temperature for 2 hours.
[k] (Z)-Stilbene is formed stereoselectively regardless of whether (Z)- or (E)-dichloroethene is used (see text).
[l] Carried out on a 0.2 mole scale at 0° to room temperature over 2 hours.

Though the cross-coupling of monohaloalkenes proceeds stereo-specifically (entry 8),[11] that of 1,2-dihaloethylenes proceeds non-stereospecifically, yielding a (Z)-alkene, the stereoselectivity being dependent on the nature of the phosphine ligand in the catalyst (entry 9).[11] Sterically hindered aryl Grignard reagents also react with ease (entry 11).[9]

The couplings are usually exothermic, and care must be taken *not* to add the phosphine–nickel catalyst to a mixture of a Grignard reagent and an organic halide, particularly in a large-scale preparation. For example the addition of a small amount of [Ni(dppp)Cl$_2$] to a mixture of vinyl chloride and 4-chlorophenylmagnesium bromide in ethyl ether at 0° led, after a few minutes' induction period, to an uncontrollable, violent reaction.

1. Department of Synthetic Chemistry, Faculty of Engineering, Kyoto University, Kyoto 606, Japan.
2. G. R. VanHecke and W. D. Horrocks, Jr., *Inorg. Chem.*, **5**, 1968 (1966).
3. M. Ogawa and G. Tanaka, *J. Chem. Soc. Jap., Ind. Chem. Sect.*, **58**, 696 (1955).
4. B. B. Elsner and H. E. Strauss, *J. Chem. Soc.*, 583 (1957).
5. K. Tamao, K. Sumitani, and M. Kumada, *J. Amer. Chem. Soc.*, **94**, 4374 (1972); R. J. P. Corriu and J. P. Masse, *J. Chem. Soc., Chem. Commun.*, 144 (1972); K. Tamao, K. Sumitani, Y. Kiso, M. Zembayashi, A. Fujioka, S. Kodama, I. Nakajima, A. Minato, and M. Kumada, *Bull. Chem. Soc. Jap.*, **49**, 1958 (1976).
6. For example, G. E. Coates, M. L. H. Green, and K. Wade, "Organometallic Compounds," Vol. 2, 3rd ed., Methuen and Co., London, 1968; W. Mowat, A. Shortland, G. Yagupsky, N. J. Hill, M. Yagupsky, and G. Wilkinson, *J. Chem. Soc., Dalton Trans.*, 533 (1972); P. S. Braterman and R. J. Cross, *Chem. Soc. Rev.*, **2**, 271 (1973).
7. K. Tamao, Y. Kiso, K. Sumitani, and M. Kumada, *J. Amer. Chem. Soc.*, **94**, 9268 (1972); Y. Kiso, K. Tamao, and M. Kumada, *J. Organometal. Chem.*, **50**, C12 (1973).
8. G. Consiglio and C. Botteghi, *Helv. Chim. Acta*, **56**, 460 (1973); Y. Kiso, K. Tamao, N. Miyake, K. Yamamoto, and M. Kumada, *Tetrahedron Lett.*, 3 (1974); T. Hayashi, M. Tajika, K. Tamao, and M. Kumada, *J. Amer. Chem. Soc.*, **98**, 3718 (1976).
9. K. Tamao, A. Minato, N. Miyake, T. Matsuda, Y. Kiso, and M. Kumada, *Chem. Lett.*, 133 (1975).
10. K. Tamao, S. Kodama, T. Nakatsuka, Y. Kiso, and M. Kumada, *J. Amer. Chem. Soc.*, **97**, 4405 (1975).
11. K. Tamao, M. Zembayashi, Y. Kiso, and M. Kumada, *J. Organometal. Chem.*, **55**, C91 (1973).
12. H. B. Kagan and T. P. Dang, *J. Amer. Chem. Soc.*, **94**, 6429 (1972).

Appendix
Chemical Abstracts Nomenclature (Collective Index Number; Registry Numbers)

Magnesium (8,9); (7439-95-4)

1-Bromobutane: Butane, 1-bromo- (8,9); (109-65-9)

1,2-Dichlorobenzene: Benzene, o-dichloro- (8); Benzene, 1,2-dichloro- (9); (95-50-1)

1-Butyl-2-chlorobenzene: Benzene, 1-butyl-2-chloro- (8,9); (15499-29-3)

1,2-Dibutylbenzene: Benzene, o-dibutyl- (8); Benzene, 1,2-dibutyl- (9); (17171-73-2)

2-Propanol (8,9); (67-63-0)

Cyclohexanone (8,9); (108-94-1)

Vinyl chloride: Ethylene, chloro- (8); Ethene, chloro- (9); (75-01-4)

Z-Stilbene: Stilbene (8); Benzene, 1,1'-(1,2-ethenediyl)bis- (9); (588-59-0); (E) (103-30-0); (Z)(645-49-8)

Ethylene, 1,2-dichloro- (8); Ethene, 1,2-dichloro- (9); (540-59-0); (E) (156-60-5); (Z) (156-59-2)

[Ni{(−)-diop}Cl₂], (−)-diop = 2,3-O-isopropylidene-2,3-dihydroxy-1,4-bis(diphenylphosphino)butane: Nickel, dichloro[[(2,2-dimethyl-1,3-dioxolane-4,5-diyl)bis(methylene)]bis[diphenylphosphine]]- (8,9); (51899-83-3); Nickel, dichloro[[(2,2-dimethyl-1,3-dioxolane-4,5-diyl)bis(methylene)]bis[diphenylphosphine]-P,P']- (8,9); (41677-72-9)

4-Chlorophenylmagnesium bromide: Magnesium, bromo(p-chlorophenyl)- (8); Magnesium, bromo(4-chlorophenyl)- (9); (873-77-8)

Ni(dmpe)Cl₂: Nickel, dichloro[ethylenebis[dimethylphosphine]]- (8); Nickel, dichloro[1,2-ethanediylbis[dimethylphosphine]-P,P']- (9); (14726-53-5)

Ni(dppe)Cl₂: Nickel, dichloro[ethylenebis[diphenylphosphine]]- (8); Nickel, dichloro[1,2-ethanediylbis[diphenylphosphine]P,P']- (SP-4-2)- (9); (14647-23-5); cis- (19978-63-3)

Ni(dppp)Cl₂: Nickel, dichloro[trimethylenebis[diphenylphosphine]]- (8); Nickel, dichloro[1,3-propanediylbis[diphenylphosphine]-P,P']- (9); (15629-92-2)

1,3-Bis(diphenylphosphino)propane: Phosphine, trimethylenebis[diphenyl- (8); Phosphine, 1,3-propanediylbis[diphenyl- (9); (6737-42-4)

[Ni(P(C₆H₅)₃)₂Cl₂]: Nickel, dichlorobis(triphenylphosphine)- (8,9); (14264-16-5)

RADICAL ANION ARYLATION: DIETHYL PHENYLPHOSPHONATE

(Phosphonic acid, phenyl-, diethyl ester)

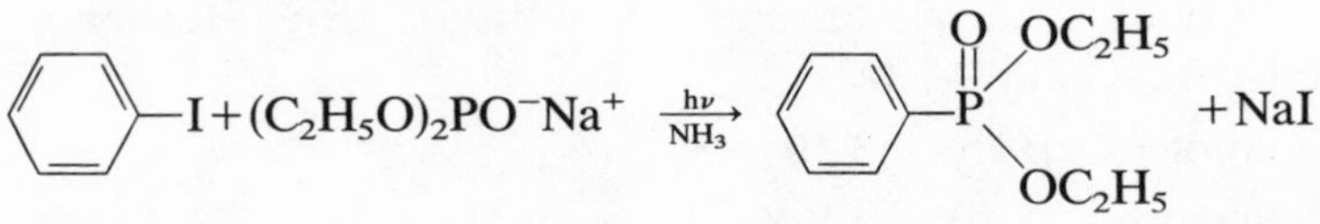

$$\text{C}_6\text{H}_5\text{—I} + (\text{C}_2\text{H}_5\text{O})_2\text{PO}^-\text{Na}^+ \xrightarrow[\text{NH}_3]{h\nu} \text{C}_6\text{H}_5\text{—P}(=\text{O})(\text{OC}_2\text{H}_5)_2 + \text{NaI}$$

Submitted by JOSEPH F. BUNNETT and ROBERT H. WEISS[1]
Checked by S. C. BUSMAN and O. L. CHAPMAN

1. Procedure

Caution! See benzene warning, p. 168.

A 2-l., three-necked, round-bottomed flask is fitted with an ammonia condenser (Note 1) with an outlet protected by a soda-lime drying tube, a dropping funnel, a nitrogen inlet, and a magnetic stirrer. As a slow stream of nitrogen is passed through the system, the condenser is charged with solid carbon dioxide and 2-propanol, the funnel is briefly removed, about a liter of liquid ammonia is added straight from a commercial cylinder (Note 2), and the funnel is replaced. Bright, freshly-pared sodium metal (11.8 g., 0.51 mole) is added, and the mixture turns blue. Diethyl phosphonate (70.4 g., 0.51 mole) (Note 3) is cautiously added dropwise to the sodium in ammonia in the manner of a titration; the endpoint is the change from blue to colorless (Note 4). Iodobenzene (52.4 g., 0.26 mole) (Note 5) is slowly added, and the solution takes on a slight yellowish tint (Note 6). The dropping funnel is replaced by a stopper, the frost is wiped off the outside of the flask with a towel dampened with acetone, and the whole system is mounted in a photochemical reactor of adequate design (Note 7). *Caution! The lamps must be shielded to prevent exposure of the eyes or skin to ultraviolet radiation.* The flask is irradiated for 1 hour (Note 8), but every 20 minutes the lamps are shut off briefly while the exterior of the flask is freed of frost by spraying it, still mounted in the reactor, with 2-propanol from a wash bottle.

After irradiation, the flask is removed from the reactor, and about 50 g. (0.62 mole) of solid ammonium nitrate is added with stirring to acidify the mixture. About 200 ml. of ethyl ether is added, nitrogen flow is stopped, the condenser is removed, and the open flask is placed on a cork ring in an operating hood and allowed to stand overnight while the ammonia evaporates. The next day 300 ml. of water (Note 9) and another 300 ml. of ethyl ether are added, the ether layer is separated, the water layer is extracted twice with ethyl ether, and the combined ether extracts are dried over anhydrous sodium sulfate. After evaporation of the ethyl ether, the residue is distilled under reduced pressure through a short Vigreux column. After a small forerun, diethyl phenylphosphonate (Note 10) is collected at 73–74° (20 μ). The yield is 50.3–56.1 g. (90.4–92.5%) (Note 11).

2. Notes

1. An ammonia condenser is an enlarged cold finger design. The interior of the "finger" is a reservoir for dry ice and 2-propanol.

2. When ammonia distilled from sodium metal is used, the yield is 3–5% greater, but use of ammonia straight from the tank is recommended because of the greater convenience.

3. Commercial diethyl phosphonate, $(C_2H_5O)_2P(O)H$, formerly known as diethyl phosphite, from Aldrich Chemical Co., Inc., was used without further purification.

4. Some white foam forms as the diethyl phosphonate is added. Because water in the ammonia consumes some of the sodium, not quite all the diethyl phosphonate is required to reach the endpoint. Excess diethyl phosphonate is deleterious.

5. Commercial iodobenzene is dried over molecular sieves. Use of ratios of potassium diethyl phosphite to iodobenzene smaller than 2:1 gives lower yields. Bromobenzene is much less reactive than iodobenzene and gives poor yields by this procedure.

6. When distilled ammonia is used, this yellow coloration vanishes as the reaction occurs.

7. The submitters used a Rayonet Model RPR-100 Photochemical Reactor, manufactured by the Southern New England Ultraviolet Company, Middletown, Connecticut, equipped with 16 Cat. No. RPR-3500A fluorescent lamps (*ca.* 24 W. each), rated to emit maximally at 350 nm. Equally good results were obtained with lamps rated to emit maximally at 300 nm. For reactions on a

0.05-mole scale excellent yields were obtained, without using the commercial photochemical reactor, by irradiating for 1 hour with two circular kitchen-type fluorescent lamps mounted on either side of the flask so as to partially encircle it.

8. The yield was not significantly improved by irradiating for 2 hours.

9. Diethyl phenylphosphonate is appreciably soluble in water. Therefore excessive amounts of water should be avoided.

10. Spectral characterization is as follows: infrared (neat) cm^{-1}: 1440 (P—C aryl), 1250 (P=O), 1020 (POC$_2$H$_5$), and 3060 (H—C aryl); proton magnetic resonance (chloroform-d) δ (multiplicity, assignment, J in Hz.): 1.3 (triplet, CH_3, $J = 7$), 4.13 (quintet, CH_2, $J = 7$), 7.33–8.06 (complex, phenyl H).

11. The submitters obtained a yield of 46 g. (83%), b.p. 90–92° (0.1 mm.).

3. Discussion

The procedure reported here is based on a reaction discovered by Bunnett and Creary,[2] and was first employed for preparative purposes by Bunnett and Traber.[3] It is attractive because of the high yield obtained, the ease of work-up, and the cleanliness of the reaction. The reaction is believed to occur by the $S_{RN}1$ mechanism, which involves radical and radical anion intermediates.[2,4] The $S_{RN}1$ arylation of other nucleophiles, especially ketone enolate ions,[5] ester enolate ions,[6] picolyl anions,[7] and arenethiolate ions,[8] has potential application in synthesis.

This procedure has been utlilized successfully with a variety of aryl iodides, but aryl bromides are much less reactive. Bromoiodobenzenes and o- and p-chloroiodobenzene give the corresponding phenylenediphosphonate esters.[2,3]

The esters of arylphosphonic acids are cleaved to the acids by HCl, HBr, or HI.[9b] Arylphosphonic dichlorides (ArPOCl$_2$) are easily converted to esters by reaction with the alcohol in pyridine solution.[10]

Other methods for synthesis of arylphosphonic acids or their derivatives fall into four main categories. First, many aromatic compounds react with PCl$_3$ under catalysis by AlCl$_3$ to form aryldichlorophosphines (ArPCl$_2$).[9a,11] These add chlorine to form aryltetrachlorophosphoranes $(ArPCl_4)$,[9d,11] which may be hydrolyzed to arylphosphonic dichlorides or arylphosphonic acids. This sequence may be employed for preparations on a large scale. It is subject to the orienting effects of substituents when applied to substituted benzenes.

Second, arylphosphonic acids may be prepared by the copper-catalyzed reactions of arenediazonium fluoborates with PCl_3 or PBr_3. This method has found wide use.[12]

Third, arylphosphonic acid derivatives have been made by organometallic reactions, such as the reaction of C_6H_5MgBr with $(C_2H_5O)_2POCl$ or of phenyllithium with phosphorodipiperididic chloride.[9c,11]

Fourth, the present procedure bears a resemblance to the photochemical reaction of aryl iodides with trialkyl phosphites, by means of which several dialkyl arylphosphonates have been prepared.[13] However, prolonged irradiation (>24 hours) in quartz vessels was employed.

1. Board of Studies in Chemistry, University of California, Santa Cruz, California 95064.
2. J. F. Bunnett and X. Creary, J. Org. Chem., **39,** 3612 (1974).
3. J. F. Bunnett and R. P. Traber, J. Org. Chem., **43,** 1867 (1978).
4. J. K. Kim and J. F. Bunnett, J. Amer. Chem. Soc., **92,** 7463, 7464 (1970).
5. R. A. Rossi and J. F. Bunnett, J. Amer. Chem. Soc., **94,** 683 (1972); J. Org. Chem., **38,** 1407 (1973); J. F. Bunnett and J. E. Sundberg, Chem. Pharm. Bull. (Jap.), **23,** 2620 (1975); J. F. Bunnett and J. E. Sundberg, J. Org. Chem., **41,** 1702 (1976).
6. X. Creary, unpublished work.
7. J. F. Bunnett and B. F. Gloor, J. Org. Chem., **39,** 382 (1974).
8. J. F. Bunnett and X. Creary, J. Org. Chem., **39,** 3173 (1974).
9. K. Sasse, "Methoden der Organischen Chemie" (Houben-Weyl), 4th ed., Band XII/1, Georg Thieme, Stuttgart, 1963, (a) p. 314; (b) p. 352; (c) p. 372; (d) p. 392.
10. T. H. Siddall, III, and C. A. Prohaska, J. Amer. Chem. Soc., **84,** 3467 (1962).
11. G. M. Kosolapoff, Org. React., **6,** 273 (1951).
12. L. D. Freedman and G. O. Doak, Chem. Rev., **57,** 479 (1957).
13. J. B. Plumb, R. Obrycki, and C. E. Griffin, J. Org. Chem., **31,** 2455 (1966); C. E. Griffin, R. B. Davison, and M. Gordon, Tetrahedron, **22,** 561 (1966); R. Obrycki and C. E. Griffin, J. Org. Chem., **33,** 632 (1968).

Appendix
Chemical Abstracts Nomenclature (Collective Index Number; Registry Numbers)

Diethyl phenylphosphonate: Phosphonic acid, phenyl-, diethyl ester (8,9); (1754-49-0)

Diethyl phosphonate: Phosphonic acid, diethyl ester (8,9); (762-04-9)

Benzene, iodo- (8,9); (591-50-4)

Acetone (8); 2-Propanone (9); (67-64-1)

Ammonium nitrate: Nitric acid, ammonium salt (8,9); (6484-52-2)

Ethyl ether (8); Ethane, 1,1'-oxybis- (9); (60-29-7)

Potassium diethyl phosphite: Phosphorous acid, diethyl ester, potassium salt (8,9); (54058-00-3)

Benzene, bromo- (8,9); (108-86-1)

C_6H_5MgBr: Magnesium, bromo phenyl (8,9); (100-58-3)

$(C_2H_5O)_2POCl$: Phosphorochloridic acid, diethyl ester (8,9); (814-49-3)

Lithium, phenyl- (8,9); (591-51-5)

Phosphorodipiperididic chloride: Phosphonic dichloride, piperidino- (8,9); (1498-56-2)

SULFIDE SYNTHESIS: BENZYL SULFIDE

(Benzene, 1,1-[thiobis(methylene)]bis-)

$$(C_6H_5CH_2)_2S_2 + [(CH_3)_2N]_3P \xrightarrow[\text{reflux}]{\text{benzene}} (C_6H_5CH_2)_2S + [(CH_3)_2N]_3PS$$

Submitted by DAVID N. HARPP and ROGER A. SMITH[1]
Checked by SUSAN A. VLADUCHICK and WILLIAM A. SHEPPARD

1. Procedure

Caution! See benzene warning, p. 168. (Note 1).

A 100-ml., one-necked, round-bottomed flask is equipped with a magnetic stirring bar and an efficient reflux condenser topped with a drying tube. The flask is charged with 3.92 g. (0.024 mole) of freshly-distilled hexamethylphosphorous triamide (Note 2), 15 ml. of dry benzene (Note 3), and 4.93 g. (0.020 mole) of benzyl disulfide (Note 4). The mixture is then refluxed for 1 hour (Notes 5 and 6), during which time it may develop a slight yellow color. After cooling to room temperature, the solution is concentrated on a rotary evaporator (Note 7), and the residue is chromatographed over 75 g. of silica gel (Note 8). After eluting with hexane–chloroform (8:2 v/v) at 3–4 ml. per minute (Note 9) and monitoring the eluant by gas chromatography (Note 10), an initial fraction containing no product is collected (typically 100 ml.). Successive fractions (*ca.* 1.45 l.) contain the product. These are combined and evaporated (Note 11), 20 ml. of absolute ethanol is added to the residue, and the mixture is heated on a steam bath for *ca.* 30 seconds to form a homogeneous solution. The solution is filtered hot through a Büchner funnel (medium-porosity fritted disk) and washed three times with 3-ml. portions of absolute ethanol. The solvent is then removed to constant weight of residue by rotary evaporation to yield 4.15–4.27 g. (96.7–99.5%) of benzyl sulfide as a white solid or as a colorless oil that crystallizes on standing (cooling on dry ice for *ca.* 1 minute may be required), m.p. 46.5–47.5°. This product is of sufficient purity for most purposes (Note 12). Further elution of the column with 1.1 l. of chloroform (same flow rate as above) and removal of solvent by rotary evaporation afford 3.94–4.06 g. (101–104%) of crude (*ca.* 95% pure) tris(dimethylamino)phosphine sulfide (Note 13).

2. Notes

1. To avoid benzene, dry toluene or acetonitrile can be used as solvents following the same procedure.

2. The hexamethylphosphorus triamide is obtained from the Eastman Kodak Company. It is distilled under reduced pressure (water aspirator) to yield a clear, colorless oil, b.p. 46–48° (7 mm.). It is then stored under nitrogen (rubber septums should be avoided as they tend to deteriorate on prolonged contact with phosphine vapors). Hexamethylphosphorus triamide that has not been freshly distilled often requires much longer reaction times. Since this chemical is an irritant, all operations with it should be performed in a well-ventilated hood. Hexamethylphosphoramide is a classified carcinogen; see *Org. Syn.*, **55**, 103 (1976). It should not be confused with hexamethylphosphorus triamide used in this preparation.

3. The benzene is dried to the extent of having been stored with type 3Å molecular sieves. Use of a greater amount of solvent slows the reaction rate. The checker flamed the reaction flask under nitrogen and ran the reaction under a nitrogen atmosphere.

4. The benzyl disulfide is obtained from the Eastman Kodak Company. Some batches, not homogeneous by gas chromatography, were recrystallized from absolute ethanol (*ca.* 2.5 ml. per g.) to yield colorless crystals, m.p. 69–71° (literature[2] m.p. 70–71.5°).

5. The temperature of the oil bath is 100–105°; the reaction mixture is *ca.* 88°.

6. After 1 hour, the reaction is complete as monitored by gas chromatographic analysis on an Hewlett-Packard F&M 5751 research chromatograph (1.85 m. × 0.313 cm. stainless-steel column of 10% Apiezon L on Chromosorb W AW/DMCS; column temperature 250°). Thin-layer chromatography was found to be of little use in monitoring the reaction, as the R_f values for benzyl disulfide and benzyl sulfide are virtually identical for a variety of solvent systems tried. Proton magnetic resonance (benzene) shows that these two compounds have coincidental chemical shifts for the benzylic protons: δ 3.4 (singlet).

A 20% excess of hexamethylphosphorus triamide is utilized to increase the reaction rate. When a 10% excess of the phosphine is used, the reaction is not quite complete (as monitored by gas chromatography) even after 3 hours. If the reaction is not allowed to go to completion, the chromatographed product will contain benzyl disulfide. Note that the desulfurization rate for hexaethylphosphorus triamide is comparable to that for hexamethylphosphorus triamide. However, the chromatographic separation following the reaction is much more efficient when hexamethylphosphorus triamide is used.

7. The pressure should be reduced gradually and cautiously to avoid the "bumping" often characteristic of benzene distillations.

8. To 75 g. of silica gel (see below) that has been freshly heated at 120–130° for 18 hours and cooled in a desiccator (see below) is added *ca.* 200 ml. of a hexane–chloroform mixture (8 : 2 v/v). The resulting slurry is stirred to remove air bubbles and is then added to a 26-mm. column (in which a flat base of glass wool and sand has been prepared) all at once, with tapping of the column. Any remaining silica gel is also added to the column by washing with more of the solvent mixture. The result is a *ca.* 28-cm. column of silica gel. A sand layer is then carefully added to the top of the column, and the solvent is drained to this sand level.

The reaction residue is placed on the column as is, and the reaction flask is rinsed with small portions of hexane–chloroform which are also loaded onto the column.

The volume of column packing used represents a length–width ratio of *ca.* 10:1. Since column efficiency improves as the length–width ratio is increased (for a given volume of packing), these approximate dimensions should be maintained. For example, when a length–width ratio of *ca.* 9:1 was applied (with the same amount of silica gel), there was not complete separation, and the yield of pure benzyl sulfide was less than optimum (93%).

The packing material is Silica Gel 60, Cat 7734 (70-230 mesh ASTM), for column chromatography, an EM Reagent of E. Merck, Darmstadt, available from Brinkmann Instruments Ltd. (Canada). Lower grades of silica gel such as the 60–120 mesh grade of BDH Chemicals Ltd. were found to be much less efficient, not effecting a complete separation of the compounds. The amount of Silica Gel 60 utilized, 75 g., or 8.5 g. of silica gel per g. of reaction mixture, was found to be the most efficient. The checkers used Mallinckrodt CC-7 Special (100–200 mesh) column packing, a flow rate of 8.5 ml. per minute and a 9:1 ratio of hexane–chloroform eluant. A lower product yield (77%) resulted.

Eighteen hours at 120–130° was found to dry various batches of Silica gel 60 to the same activity. The silica gel was cooled in a desiccator over 8-mesh Drierite available from Anachemia Chemicals Ltd. This procedure should be performed so that the silica gel is used immediately after cooling. Otherwise, the activity will decrease with time even if the silica is left in the desiccator.

The solvent mixture of 8 parts hexane to 2 parts chloroform (v/v) was found to be the most effective, with increased concentrations of chloroform not effecting a complete separation. The two solvents should therefore be mixed thoroughly. Chloroform (A.C.S. grade) and an isomeric mixture of hexanes (A.C.S. grade) supplied by Fisher Scientific Company are quite suitable.

9. Increased flow rate results in a poorer chromatographic resolution.

10. The chromatography may be monitored by collection of 50-ml. fractions analyzed by gas chromatography (same conditions as in Note 6).

11. The residue will be a white solid or colorless oil.

12. The solidified residue is homogeneous by gas chromatography, and its infrared and nuclear magnetic resonance spectra and gas chromatogram are identical to those of recrystallized authentic benzyl sulfide. The product may be recrystallized in *ca.* 25 ml. of absolute ethanol to yield colorless plates, m.p. 47–48° (literature[3] m.p. 50°), which give a satisfactory combustion analysis.

13. This material has infrared and nuclear magnetic resonance spectra identical to those of an authentic sample prepared by the procedure of Stuebe and Lankelma.[4]

3. Discussion

The disulfide linkage is found in a large number of natural products,[5] and chemical manipulations of this functionality are of considerable interest.[6] In contrast to desulfurization methods utilizing various phosphines[7] and phosphites,[8] aminophosphines have proven to be efficient desulfurizing agents for a variety of alkyl, aralkyl, alicyclic, and certain diaryl disulfides.[9] In addition, the reaction conditions are mild enough to be compatible with a variety of common functional groups. Hence, the desulfurization procedure given here for benzyl disulfide merely demonstrates the practibility of this method for which the synthetic scope and mechanism have already been developed.[9]

1. Department of Chemistry, McGill University, P.O. Box 6070, Montreal, Quebec, Canada H3C 3G1.
2. F. H. McMillan and J. A. King, *J. Amer. Chem. Soc.*, **70**, 4143 (1948).
3. C. Forst, *Justus Liebigs Ann. Chem.*, **178**, 370 (1875).
4. C. Stuebe and H. P. Lankelma, *J. Amer. Chem. Soc.*, **78**, 976 (1956).
5. (a) L. Young and G. A. Maw, "The Metabolism of Sulfur Compounds," John Wiley & Sons, New York, N. Y., 1958; (b) R. F. Steiner, "The Chemical Foundations of Molecular Biology," D. Van Nostrand Company, Princeton, N. J., 1965; (c) K. G. Stern and A. White, *J. Biol. Chem.*, **117**, 95 (1937); (d) K. Jost, V. Debabov, H. Nesvadba, and J. Rudinger, *Collect. Czech. Chem. Commun.*, **29**, 419 (1964).
6. (a) W. A. Pryor and K. Smith, *J. Amer. Chem. Soc.*, **92**, 2731 (1970); (b) J. L. Kice, "Mechanisms of Reactions of Sulfur Compounds," Vol. 3, John Wiley & Sons, New York, N. Y., 1968, p. 91; (c) N. Kharasch and A. J. Parker, *Q. Rep. Sulfur Chem.*, **1**, 285 (1966); (d) N. Kharasch, Ed., "Organic Sulfur Compounds," Vol. 1, Pergamon Press, New York, N. Y., 1961; (e) A. J. Parker and N. Kharasch, *Chem. Rev.*, **59**, 583, 589 (1959).
7. (a) A. Schönberg, *Chem. Ber.*, **68**, 163 (1935); (b) A. Schönberg and M. Barakat, *J. Chem. Soc.*, 892 (1949); (c) F. Challenger and D. Greenwood, *J. Chem. Soc.*, 26 (1950); (d) C. Moore and B. Trego, *Tetrahedron*, **18**, 205 (1962).

8. (a) H. Jacobson, R. Harvey, and E. V. Jensen, *J. Amer. Chem. Soc.*, **77**, 6064 (1955); (b) A. Poshkus and J. Herweh, *J. Amer. Chem. Soc.*, **79**, 4245 (1957); (c) C. Walling and R. Rabinowitz, *J. Amer. Chem. Soc.*, **79**, 5326 (1957); (d) C. Walling and R. Rabinowitz, *J. Amer. Chem. Soc.*, **81**, 1243 (1959); (e) C. Moore and B. Trego, *J. Chem. Soc.*, 4205 (1962); (f) R. Harvey, H. Jacobson, and E. Jensen, *J. Amer. Chem. Soc.*, **85**, 1618 (1963); (g) K. Pilgram, D. Phillips, and F. Korte, *J. Org. Chem.*, **29**, 1844 (1964); (h) K. Pilgram and F. Korte, *Tetrahedron*, **21**, 203 (1965); (i) A. J. Kirby, *Tetrahedron*, **22**, 3001 (1966); (j) R. S. Davidson, *J. Chem. Soc.*, C. 2131 (1967).
9. D. N. Harpp and J. G. Gleason, *J. Amer. Chem. Soc.*, **93**, 2437 (1971).

Appendix
Chemical Abstracts Nomenclature (Collective Index Number; Registry Numbers)

Benzyl sulfide (8); Benzene, 1,1'-[thiobis(methylene)]bis- (9); (538-74-9)

Phosphorus triamide, hexamethyl- (8,9); (1608-26-0)

Benzyl disulfide (8); Disulfide, dibenzyl (8); Disulfide, bis(phenylmethyl)- (9); (150-60-7)

Tris(dimethylamino)phosphine sulfide: Phosphorothioic triamide, hexamethyl- (8,9); (3732-82-9)

Phosphorus triamide, hexaethyl- (8,9); (2283-11-6)

SULFIDE SYNTHESIS IN PREPARATION OF DIALKYL AND ALKYL ARYL SULFIDES: NEOPENTYL PHENYL SULFIDE
(Benzene, [(2,2-dimethylpropyl)thio]-)

$$(CH_3)_3CCH_2Br + C_6H_5\overset{-}{S}\overset{+}{Na} \xrightarrow[H_2O]{(C_4H_9)_3C_{16}H_{33}\overset{+}{P}\overset{-}{Br}} C_6H_5SCH_2C(CH_3)_3 + NaBr$$

Submitted by D. Landini and F. Rolla[1]
Checked by Ronald L. Sobczak and S. Masamune

1. Procedure

A 100-ml., two-necked flask is fitted with a reflux condenser, a gas inlet, and a magnetic stirrer. The flask is charged with 12.0 ml. (15.1 g., 0.1 mole) of 1-bromo-2,2-dimethylpropane (Note 1), aqueous sodium benzenethiolate (0.1 mole) (Note 2), and 1.67 g. (0.0033 mole) of tributylhexadecylphosphonium bromide (Notes 3 and 4). This mixture is heated at 70° with vigorous stirring under nitrogen (Note 5) for 3.5 hours (Note 6). After the mixture has cooled to room temperature, the organic layer is separated and the aqueous phase is extracted with two 20-ml. portions of ethyl ether. The combined organic phases are washed with 20 ml. of aqueous 10% sodium chloride and dried over calcium chloride. After removal of the solvent, the resulting residual oil is distilled through a 10-cm. Vigreux column to give 14.1–15.3 g. (78–85%) of colorless neopentyl phenyl sulfide (Note 7), b.p. 85–87° (5 mm.), 96–98° (8 mm.); n^{24}D 1.5365 (Note 8).

2. Notes

1. 1-Bromo-2,2-dimethylpropane (neopentyl bromide) was obtained from Fluka A G or Tridom Chemical Inc.

2. Aqueous sodium benzenethiolate was prepared by adding 10.2 ml. (11.0 g., 0.1 mole) of commercial benzenethiol (listed as thiophenol by Aldrich Chemical Company, Inc. and Tridom Chemical Inc.) to an ice-cold solution of 4.0 g. of sodium hydroxide in 25 ml. of water.

3. The tributylhexadecylphosphonium bromide was prepared by heating 0.1 mole of 1-bromohexadecane and 0.1 mole of tributylphosphine at 60–70° for three days, according to Starks' procedure.[2] The product, while hot, was poured into 300 ml. of hexane and the mixture was stirred for 15 minutes. After cooling of the mixture to 0°, a solid product crystallized and was filtered on a Büchner funnel and dried under reduced pressure. It had a melting point of 54–56° (84%).

4. When the reaction was carried out using 0.033 mole equivalent of tricaprylylmethylammonium chloride (aliquat 336), obtained from General Mills Company, Chemical Division, Kankakee, Illinois, as catalyst, the reaction required about 10 hours for its completion.

5. The nitrogen flow must be as slow as possible to avoid loss of 1-bromo-2,2-dimethylpropane.

6. The reaction time depends on the concentration of the catalyst; e.g., with 0.1 and 0.01 mole equivalents of phosphonium salt, the reaction required 1 and 10 hours, respectively.

7. The catalyst could be recovered (80–90%) from the distillation residue, which also contained some neopentyl phenyl sulfide

and diphenyl disulfide. These products were eliminated from the residue by column chromatography on silica (8 g. for 1 g. of phosphonium salt; eluant, ethyl ether). Extraction of the silica with two 25-ml. portions of boiling ethanol and evaporation of the solvent afforded the phosphonium salt, m.p. 48–51°. This material could be reused without further purification.

8. The product showed the following proton magnetic resonance spectrum (chloroform-d) δ (multiplicity, number of protons, assignment): 1.03 [singlet, 9, $(CH_3)_3C$], 2.88 (singlet, 2, CH_2), 7.02–7.52 (multiplet, 5, C_6H_5).

3. Discussion

This procedure[3] is an example of a simple and general method for preparation of primary and secondary dialkyl and alkyl aryl thioethers via alkylation of sodium sulfide or sodium alkyl- or arylthiolates with alkyl chlorides or bromides. The method is an

TABLE I

PREPARATION OF DIALKYL AND ALKYL PHENYL SULFIDES

Alkyl Halide	Nucleophile	Catalyst (mole equiv- alent)	Temp- erature (°C)	Time (min- utes)	Yield of Sulfide[a] (%)
1-Chlorooctane	Na_2S[b]	0.1	70	40	91
2-Chlorooctane	Na_2S[b]	0.1	70	300	90
1-Bromooctane	Na_2S[b]	0.1	70	20	91
2-Bromooctane	Na_2S[b]	0.1	70	80	91
Neopentylbromide	Na_2S[b]	0.1	70	500	81[c]
1-Chlorooctane	C_2H_5SNa[d]	0.033	40	40	90
2-Chlorooctane	C_2H_5SNa[d]	0.033	70	250	88
1-Bromooctane	C_2H_5SNa[d]	0.033	40	15	91
2-Bromooctane	C_2H_5SNa[d]	0.033	70	120	89
1-Chlorooctane	C_6H_5SNa[d]	0.033	40	30	92
2-Chlorooctane	C_6H_5SNa[d]	0.033	70	180	90
1-Bromooctane	C_6H_5SNa[d]	0.033	40	10	91
2-Bromooctane	C_6H_5SNa[d]	0.033	70	60	90

[a] Isolated products.
[b] Mole ratio of Na_2S to alkyl halide is 0.6.
[c] Reaction carried out under nitrogen.
[d] Mole ratio of sodium salt to alkyl halide is 1.

example of phase-transfer catalysis characterized by mild reaction conditions, high yields, and simple work-up procedure.

Dineopentyl and neopentyl phenyl sulfides are obtained from 1-bromo-2,2-dimethylpropane. Some other examples are given in Table I.

Neopentyl sulfides have been prepared by alkylation of sodium sulfide with neopentyl tosylate in high-boiling polar solvents,[4,5] or in low yields by reduction of alkyl 2,2-dimethylpropanethioate with the combination of lithium aluminum hydride and a large excess of boron trifluoride-etherate.[6]

1. Centro C.N.R. e Istituto di Chimica Industriale dell'Universita', Via C. Golgi 19, Milano 20133, Italy.
2. C. M. Starks, J. Amer. Chem. Soc., **93**, 195 (1971).
3. D. Landini and F. Rolla, Synthesis, 565 (1974).
4. F. G. Bordwell, B. M. Pitt, and M. Knell, J. Amer. Chem. Soc., **73**, 5004 (1951).
5. W. E. Parham and L. D. Edwards, J. Org. Chem., **33**, 4150 (1968).
6. E. L. Eliel and R. A. Daignault, J. Org. Chem., **29**, 1630 (1964).

Appendix
Chemical Abstracts Nomenclature; (Collective Index Number; Registry Numbers)

Sulfide, neopentyl phenyl (8); Benzene, [(2,2-dimethylpropyl)thio]- (9); (7210-80-2)

Propane, 1-bromo-2,2-dimethyl- (8,9); (630-17-1)

Sodium benzenethiolate: Benzenethiol, sodium salt (8,9); (930-69-8)

Tributylhexadecylphosphonium bromide: Phosphonium, tributyl-hexadecyl-, bromide (8,9); (14937-45-2)

Benzenethiol (8,9); (108-98-5)

Hexadecane, 1-bromo- (8,9); (112-82-3)

Phosphine, tributyl- (8,9); (998-40-3)

Tricaprylylmethylammonium chloride: Ammonium, methyl-trioctanoyl-, chloride (8); 1-Octanaminium, N-methyl-1-oxo-N,N-bis(1-oxooctyl)-, chloride (9); (13275-89-3)

Disulfide, diphenyl- (8,9); (822-33-7)

Dineopentyl sulfide: Neopentyl sulfide (8); Propane, 1,1'-thiobis-[2,2-dimethyl- (9); (51616-83-2)

Neopentyl tosylate: p-Toluenesulfinic acid, neopentyl ester (8);
Benzenesulfinic acid, 4-methyl-2,2-dimethylpropyl ester (9);
(13146-08-2)
Octane, 1-chloro- (8,9); (111-85-3)
Octane, 2-chloro- (8,9); (628-61-5)
Octane, 1-bromo- (8,9); (111-83-1)
Octane, 2-bromo- (8,9); (557-35-7)

SULFIDE SYNTHESIS IN PREPARATION OF UNSYMMETRICAL DIALKYL DISULFIDES: sec-BUTYL ISOPROPYL DISULFIDE
(Disulfide, 1-methylethyl 1-methylpropyl)

A. $(CH_3)_2CHBr + Na_2S_2O_3 \cdot 5H_2O \xrightarrow[\text{reflux}]{CH_3OH-H_2O} (CH_3)_2CH-S-SO_3Na + NaBr$

1

B. $CH_3CH_2CH(CH_3)-SH + NaOH \xrightarrow[25°]{H_2O} CH_3CH_2CH(CH_3)-SNa$

2

C. $1 + 2 \xrightarrow[0-5°]{H_2O} (CH_3)_2CH-S-S-CH(CH_3)CH_2CH_3$

3

Submitted by M. E. Alonso[1] and H. Aragona
Checked by Linda D.-L. Lu and S. Masamune

1. Procedure

Caution! This procedure should be carried out in an efficient hood to prevent exposure to alkane thiols.

A. *Sodium isopropylthiosulfate* (**1**). A 5-l., three-necked, round-bottomed flask is equipped with a condenser, a 300-ml. dropping funnel, and a mechanical stirrer. The flask is charged with 123 g. (1 mole) of freshly distilled 2-bromopropane (Note 1) in 1.2 l. of methanol. Water is added slowly with stirring until a slight turbidity develops (Note 2). The stirred mixture is heated to reflux, and 310 g. (1.25 mole) of sodium thiosulfate pentahydrate (Note 3) in 250 ml. of water is added over a period of 30 minutes. The slightly yellow tinted solution is heated for an additional 3.5–4.0 hours and then allowed to cool to room temperature. The alcohol is removed on a rotary evaporator, and the remaining milky solution is diluted with water to a total volume of about 1.2 l. The solution is extracted twice with hexane (Note 4), and the organic layer discarded. The aqueous solution of crude thiosulfate **1** is cooled to 0° and stored.

B. *Sodium 2-butanethiolate* (**2**). A 500-ml., three-necked, round-bottomed flask is equipped with a dropping funnel, a mechanical stirrer, and a gas inlet. The flask is charged with 40 g. (1 mole) of sodium hydroxide in 100 ml. of water. Under an atmosphere of argon (Note 5), 90 g. (1 mole) of 2-butanethiol (Note 6) is added dropwise over a 2-hour period with rapid stirring at room temperature (Note 7). This thiolate solution **2** becomes very viscous toward the end of the addition; it is diluted with 30 ml. of water and then cooled to 0°.

C. sec-*Butyl isopropyl disulfide* (**3**). To a 3-l., three-necked, round-bottomed flask are attached a dropping funnel, a thermometer, and a mechanical stirrer. The flask is charged with the crude thiosulfate solution **1** and cooled to 0° with the aid of an ice–salt bath. The cold thiolate solution **2** is added rapidly with vigorous stirring for 3 minutes. Then 200 ml. of aqueous saturated sodium chloride is added (Note 8), and the mixture is warmed to 5°. Stirring is stopped after 10 minutes, counted from the start of the addition of the aqueous sodium chloride. The crude disulfide **3,** which separates as an oil, is subsequently removed, and the aqueous layer is extracted twice with 250-ml. portions of ethyl ether. The extracts are combined with the oil, washed twice with 150-ml. portions of water, and then dried briefly over granular calcium sulfate. The drying agent is removed by filtration through a glass-wool plug. Removal of the solvent leaves 125–133 g. of crude disulfide **3** (Note 9). Distillation of this material at 44.5–45° (1.25 mm.) gives 118–123 g. (73–75%) of pure disulfide **3** (Note 10).

2. Notes

1. The submitters used 2-bromopropane available from Aldrich Chemical Company, Inc. The checkers purchased the reagent from J. T. Baker Chemical Company.

2. The appearance of turbidity indicates saturation of alkyl halide. In this way both sodium thiosulfate and 2-bromopropane are nearly in a one-phase system, thus shortening significantly the heating period. Furthermore, the competitive hydrogen bromide elimination and the ensuing acid-promoted decomposition of thiosulfate into sulfur and sulfur dioxide are minimized. the checkers added 300 ml. of water over a period of 90 minutes.

3. The submitters used sodium thiosulfate pentahydrate supplied by E. Merck A G, Darmstadt, and the checkers used A.C.S. reagent-grade material available from Fisher Scientific Company.

4. This extraction is intended to remove traces of unreacted alkyl halide that might compete for the thiolate in the nucleophilic substitution (Step B).

5. The submitters performed this step in air. The checkers found that use of an inert atmosphere resulted in a somewhat improved yield.

6. This reagent was purchased from Aldrich Chemical Company, Inc.

7. The submitters observed the separation of a pasty solid at this stage and added four 10-ml. portions of water during the addition of the thiol, then dissolved the entire solid with approximately 240 ml. of water. Sometimes, at this point, as much as 10% of the added thiol separated out as a floating oil. The presence of this thiol affects the course of the reaction to yield symmetrical disulfides.[2] In this case the organic layer should be separated, then added dropwise to an equivalent amount of sodium hydroxide dissolved in a minimum amount of water, and mixed with the original thiolate solution.

8. The submitters found that the addition of sodium chloride facilitated the separation of the insoluble disulfide.

9. This material contained no less than 90% of disulfide 3 according to gas chromatographic analysis (1.5-m. 5% SE-30 column).

10. The distilled disulfide 3 has the following proton magnetic resonance spectrum (chloroform-d) δ (multiplicity, number of protons, assignment, coupling constant J in Hz.): 1.00 (triplet, 3, CH_3CH_2, $J = 7.0$), 1.31 [doublet, 9, $(CH_3)_2CH$ and CH_3CH, $J = 7.0$], 1.62, (multiplet, 2, CH_2), 2.86 (sextuplet, 1, CH_3CHCH_2, $J = 7.0$), 3.02 (septuplet, 1, CH_3CHCH_3, $J = 7.0$).

3. Discussion

Unsymmetrical dialkyl disulfides[3] can be prepared by several methods, all of which involve the attack of a nucleophilic form of sulfur on a sulfur atom bearing a suitable leaving group. Three procedures appear to be generally applicable. First, the reaction of an N-(alkylthio)- or N-(arylthio)phthalimide with thiols[4] gives unsymmetrical disulfides in good yield. However, the synthesis of the thiophthalimide[5] requires the corresponding sulfenyl chloride, which is rather unstable and undergoes undesirable side reactions when α-protons are available.[6] Second, addition of a thiol to diethyl azodicarboxylate gives the corresponding adduct, which reacts with a thiol to give unsymmetrical disulfides in high yield.[7] The adduct formation, however, is severely suppressed by steric hindrance in the alkyl portion of the thiol. Secondary and tertiary thiols are normally unreactive.[8a] Third, the reaction of sodium alkylthiosulfates[10] with thiolates provides unhindered mixed disulfides in low to moderate yields[9] and hindered compounds[8] in yields of 6–10%. The availability and low cost of starting materials and the expeditious processes involved provided a reasonable basis for modifying the existing procedure[9] to improve its applicability even to bulky unsymmetrical disulfides. Table I shows boiling points and distillate composition of a number of mixed disulfides prepared in up to 80% yield by the method presented here. The

TABLE I
UNSYMMETRICAL DIALKYL DISULFIDES (RSSR') PREPARED BY THE THIOSULFATE PROCEDURE

R	R'	b.p.° (mm.)	Yield (%)	Composition (%)[a]		
				RSSR'	RSSR	R'SSR'
$(CH_3)_2CHCH_2$—	$(CH_3)_2CH$—	33–34 (0.15)	60	92	6	2
$(CH_3)_2CHCH_2$—	CH_3CH_2—	51 (3)	62	98	—	1
$(CH_3)_2CHCH_2$—	$CH_2{=}CHCH_2$—	57 (1.5)	81	96	2	1.5
$(CH_3)_2CH$—	CH_3CH_2—	49 (9)	60	98	<1	b
$(CH_3)_2CH$—	CH_3—	37–38 (11)	72	98.5	<1	—
$C_2H_5CH(CH_3)$—	$(CH_3)_2CH$—	40–41 (0.2)	73	96.5	3	b
$C_2H_5CH(CH_3)$—	CH_3CH_2—	47 (1.5)	64	98.5	<1	<1
$(CH_3)_3C$—	CH_3CH_2—	43–44 (5)	52	97	<1	2

[a] Gas–liquid chromatographic analysis with a 1.5-m. 5% SE-30 column at 135°.
[b] Could not be separated by gas chromatography.

side reactions of base-catalyzed disproportionation[11] and poly-sulfide formation[9] appear to be minimized.

1. Centro de Petróleo y Química, Instituto Venezolano de Investigaciones Científicas, I.V.I.C., Apartado 1827 Caracas 101, Venezuela.
2. A. J. Parker and N. Kharasch, *Chem. Rev.*, **59**, 583 (1959).
3. A. Schöberl and A. Wagner, "Methoden der Organischen Chemie" (Houben–Weyl), Vol. 9, 4th ed., E. Müller, Ed., G. Thieme Verlag, Stuttgart, 1955, pp. 72–73; C. G. Moore and M. Porter, *J. Chem. Soc.*, 2890 (1958); L. Field, T. C. Owen, R. R. Crenshaw, and A. W. Bryan, *J. Amer. Chem. Soc.*, **83**, 4414 (1961); T. F. Parsons, J. D. Buckman, D. E. Pearson, and L. Field, *J. Org. Chem.*, **30**, 1923 (1965), and references cited therein; V. I. Dronov and N. V. Pokaneshchikova, *Zh. Org. Khim.*, **6**, 2225 (1970); R. H. Cragg, J. P. N. Husband, and A. F. Weston, *Chem. Commun.*, 1701 (1970); L. Field, *Synthesis*, 101 (1972); J. Meijer and P. Vermeer, *Rec. Trav. Chim. Pays-Bas*, **93**, 242 (1974).
4. K. S. Boustany and A. B. Sullivan, *Tetrahedron Lett.*, 3547 (1970); D. N. Harpp, D. K. Ash, T. G. Back, J. G. Gleason, B. A. Orwig, W. F. Van Horn, and J. P. Snyder, *Tetrahedron Lett.*, 3551 (1970).
5. E. Kühle, *Synthesis*, 617 (1971); M. Fukurawa, T. Suda, A. Tsukamoto, and S. Hayashi, *Synthesis*, 165 (1975).
6. M. Behforouz and J. E. Kerwood, *J. Org. Chem.*, **34**, 51 (1969).
7. T. Mukaiyama and K. Takahashi, *Tetrahedron Lett.*, 5907 (1968).
8. (a) M. E. Alonso, unpublished results; (b) H. E. Wijers, H. Boelens, A. Van der Gen, and L. Brandsma, *Rec. Trav. Chim. Pays-Bas*, **88**, 519 (1969).
9. B. Milligan and J. M. Swan, *J. Chem. Soc.*, 6008 (1963); B. Milligan, B. Saville, and J. M. Swan, *J. Chem. Soc.*, 4850 (1961).
10. H. B. Footner and S. Smiles, *J. Chem. Soc.*, **127**, 2887 (1925).
11. M. Calvin, "Mercaptans and Disulfides," Oak Ridge, Tenn., 1954, U.S. Atomic Energy Report UCRL-2438; A. P. Ryle and F. Sanger, *Biochem. J.*, **60**, 535 (1955); J. L. Kice and G. E. Ekman, *J. Org. Chem.*, **40**, 711 (1975); J. P. Danehy, "The Chemistry of Organic Sulfur Compounds," Vol. 2, N. Kharasch and C. Y. Meyers, Ed., Pergamon Press, New York, N. Y., 1966, 337.

Appendix
Chemical Abstracts Nomenclature (Collective Index Number; Registry Numbers)

Disulfide, *sec*-butyl isopropyl (8); Disulfide, 1-methylethyl 1-methylpropyl (9); (−)

Sodium isopropylthiosulfate: Thiosulfuric acid, *S*-isopropyl ester, sodium salt (8); Thiosulfuric acid, *S*-(1-methylethyl) ester, sodium salt (9); (26726-19-2)

Propane, 2-bromo- (8,9); (75-26-3)

Sodium thiosulfate pentahydrate: Thiosulfuric acid, disodium salt, pentahydrate (8,9); (10102-17-7)

2-Butanethiol (8,9); (513-53-1)

Diethyl azodicarboxylate: Formic acid, azodi-, diethyl ester (8); Diazenedicarboxylic acid, diethyl ester (9); (1972-28-7)

3-TRIMETHYLSILYL-3-BUTEN-2-ONE: A MICHAEL ACCEPTOR

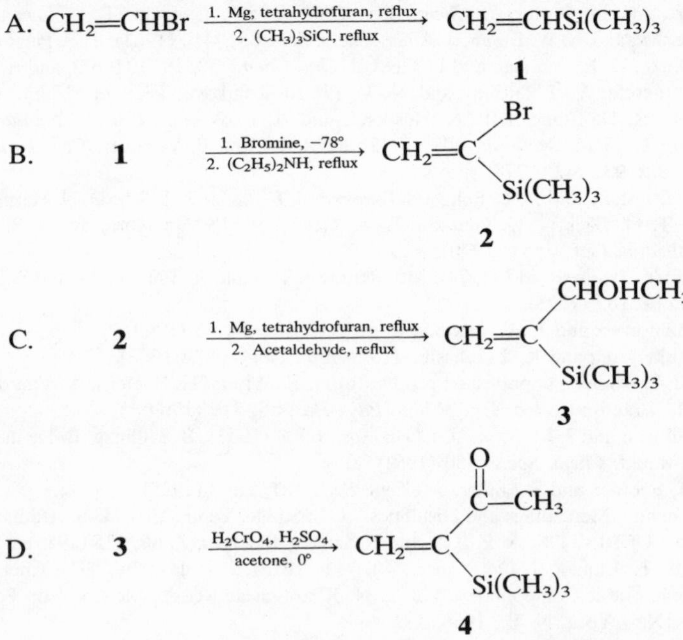

A. $CH_2=CHBr$ $\xrightarrow[\text{2. }(CH_3)_3SiCl,\text{ reflux}]{\text{1. Mg, tetrahydrofuran, reflux}}$ $CH_2=CHSi(CH_3)_3$

1

B.   **1** $\xrightarrow[\text{2. }(C_2H_5)_2NH,\text{ reflux}]{\text{1. Bromine, }-78°}$ $CH_2=C\begin{smallmatrix}Br\\[2pt]\\Si(CH_3)_3\end{smallmatrix}$

2

C.   **2** $\xrightarrow[\text{2. Acetaldehyde, reflux}]{\text{1. Mg, tetrahydrofuran, reflux}}$ $CH_2=C\begin{smallmatrix}CHOHCH_3\\[2pt]\\Si(CH_3)_3\end{smallmatrix}$

3

D.   **3** $\xrightarrow[\text{acetone, }0°]{H_2CrO_4,\ H_2SO_4}$ $CH_2=C\begin{smallmatrix}\overset{O}{\overset{\|}{C}}-CH_3\\[2pt]\\Si(CH_3)_3\end{smallmatrix}$

4

Submitted by ROBERT K. BOECKMAN, JR.,[1] DAVID M. BLUM,[1] BRUCE GANEM,[2] and NEIL HALVEY[2]
Checked by WILLIAM R. BAKER and ROBERT M. COATES

1. Procedure

A. *Vinyltrimethylsilane* (**1**). A 2-l., three-necked, round-bottomed flask is fitted with a mechanical stirrer, a reflux condenser, and a 500-ml. pressure-equalizing dropping funnel (Note 1). The flask is charged with 26.4 g. (1.1 mole) of magnesium turnings and 800 ml. of dry tetrahydrofuran (Note 2). A solution of 107 g. (72 ml., 1.0 mole) of vinyl bromide (Note 3) in 200 ml. of tetrahydrofuran is placed in the addition funnel and slowly added dropwise to the reaction vessel. After the reaction has begun (Note 4), the addition rate is regulated to maintain a gentle reflux during the remainder of the addition period. The mixture is heated at reflux for an additional hour, and a solution of 108 g. (1.0 mole) of chlorotrimethylsilane (Note 5) in 100 ml. of tetrahydrofuran is added dropwise while the reaction is maintained at reflux with continued heating and stirring (Note 6). The suspension is stirred for another 2 hours under reflux, then cooled to room temperature and stirred overnight.

The condenser and dropping funnel are removed, and the flask is equipped for distillation with a 30.5-cm. Vigreux column. The distillate (b.p. 60–65°) is collected, transferred to a separatory funnel, and washed with 10–20 100-ml. portions of water (Note 7). The remaining colorless liquid, which amounts to 67–78 g. (67–78%), is predominantly silane **1**, but contains small amounts of tetrahydrofuran (Note 8).

B. 1-(*Bromovinyl*)*trimethylsilane* (**2**). A 1-l., three-necked, round-bottomed flask is equipped with a mechanical stirrer and a 250-ml. dropping funnel. The flask is charged with 89.8 g. (approximately 0.90 mole, Note 8) of the silane **1**. The contents of the flask are stirred and cooled to −78°, and 168 g. (1.06 mole) of bromine is added dropwise over *ca.* 1 hour. The cooling bath is removed, and the red viscous mixture is warmed to room temperature. The flask is fitted with an efficient, water-cooled condenser, and 600 ml. (425 g., 5.82 moles) of diethylamine (Note 9) is cautiously (Note 10) added with continued stirring. After the addition is complete, the reaction mixture is heated at reflux for 12 hours, during which time a precipitate of diethylamine hydrochloride forms. The salts are separated from the cooled suspension by filtration and washed with several 300-ml. portions of ethyl ether. The ether filtrate is carefully washed, first with 100-ml. portions of 10% aqueous hydrochloric acid until the aqueous layer remains acidic (pH *ca.* 2), and then with 100 ml. of water and 200 ml. of saturated aqueous sodium chloride. The ether solution is dried with anhydrous magnesium sulfate, concentrated with a rotary evaporator, and distilled under reduced pressure through a 20.3-cm. Vigreux column, affording 104–110 g. (65–68%) of silane **2**, b.p. 72–75° (120 mm.) (Note 11).

C. 3-*Trimethylsilyl*-3-*buten*-2-*ol* (**3**). A 500-ml., three-necked, round-bottomed flask is equipped with two 30.5-cm. Liebig condensers connected in series, a pressure-equalizing dropping funnel, and a magnetic stirrer. The flask is charged with 9.2 g. (0.38 mole) of magnesium turnings and 100 ml. of tetrahydrofuran (Notes 1 and 2). About 2 g. of 1,2-dibromoethane is added to initiate the formation of the Grignard reagent. When the supernatant solution becomes warm and begins to reflux from reduction of 1,2-dibromoethane, a solution of 50 g. (0.28 mole) of silane 2 in 75 ml. of tetrahydrofuran is added dropwise to the stirred mixture at a rate that maintains gentle reflux. After the addition is complete, the reaction mixture is kept at reflux for an additional hour. Then freshly-distilled acetaldehyde (25.0 g., 0.57 mole) is introduced. The temperature is maintained at reflux, and stirring is continued throughout the addition and for an additional hour. The flask is then fitted with a distillation head and heated until a distillate amounting to approximately 100 ml. has been collected. The reaction mixture is cooled (ice-water bath) and stirred, diluted with 100 ml. of ethyl ether, and hydrolyzed by addition of enough saturated ammonium chloride (approximately 50 ml.) to dissolve the thick, sticky precipitate. The salts are filtered and washed with ethyl ether, and the aqueous layer of the filtrate is extracted with three 150-ml. portions of ethyl ether. The combined ether layers are washed with saturated aqueous sodium chloride, dried over anhydrous magnesium sulfate, and concentrated by distillation at atmospheric pressure (Note 12), giving 48–55 g. of crude butenol 3 as a liquid that is used in the next step without further purification.

D. 3-*Trimethylsilyl-3-buten*-2-*one* (4). A solution of 55 g. of crude butenol 3 in 100 ml. of acetone is placed in a 500-ml., three-necked, round-bottomed flask equipped with a mechanical stirrer and a 250-ml. dropping funnel. The reaction vessel is immersed in an ice-water bath, and 95 ml. of an aqueous solution containing chromic acid and sulfuric acid (Note 13) is added to the stirred acetone solution. After completion of the addition, isopropyl alcohol is added to the reaction mixture until a green endpoint is reached, indicating consumption of the excess oxidant. The contents are poured into 450 ml. of ethyl ether, 300 ml. of water are added, and the aqueous layer is saturated with sodium chloride. The layers are separated, and the aqueous solution is extracted with five 150-ml. portions of ethyl ether. The combined ether solutions are washed with two 150-ml. portions of saturated aqueous sodium chloride, dried with anhydrous magnesium sulfate, and concentrated by distillation at atmospheric pressure through a 30.5-cm. Vigreux column. Continued distillation under reduced pressure gives, after separation of a low boiling forerun, 14.7–15 g. (37–38%) of butenone **4** as a pale yellow liquid: b.p. 98–103° (100 mm.) (Notes 14 and 15).

2. Notes

1. The apparatus is flamed dry under an argon atmosphere and maintained under argon during the reaction.

2. The submitters purified the tetrahydrofuran by distillation from lithium aluminum hydride. The checkers used tetrahydrofuran that had been distilled from the sodium ketyl of benzophenone. (*Caution!* See *Org. Syn.*, Coll. Vol. **5**, 976 (1973), *for a warning regarding the purification of tetrahydrofuran.*)

3. Gaseous vinyl bromide is condensed in a 500-ml. flask cooled in an acetone–dry ice bath and then diluted with 200 ml. of tetrahydrofuran. The submitters used vinyl bromide supplied by J. T. Baker Chemical Company; the checkers purchased this reagent from Linde Specialty Gases. Vinyl bromide is also available from Aldrich Chemical Company, Inc.

4. Approximately 70 ml. is added over a 20-minute period before formation of the Grignard reagent begins. The total addition time is *ca.* 1 hour.

5. The submitters used practical-grade chlorotrimethylsilane purchased from PCR, Inc., which was distilled before use. Chlorotrimethylsilane from Aldrich Chemical Company, Inc., was employed by the checkers, both with and without prior distillation. Approximately the same yield was obtained in either case.

6. A white precipitate, presumed to be magnesium salts, is deposited as the solution of chlorotrimethylsilane is added.

7. The layers must be allowed to separate completely to avoid sizable mechanical losses. The submitters used 10 water extractions, whereas the checkers continued the extractions until the organic layer reached an approximately constant weight.

8. The proton magnetic resonance spectrum of the product obtained by the checkers revealed the presence of $6 \pm 2\%$ of tetrahydrofuran.

9. Diethylamine is available from Aldrich Chemical Company, Inc.

10. The reaction of the excess bromine with diethylamine is exothermic; consequently it may be necessary to moderate the reaction by cooling with an ice-water bath during the early stages of the addition.

11. Proton magnetic resonance spectrum (chloroform-d) δ (multiplicity, number of protons, assignment, coupling constant J in Hz.): 0.16 (singlet, 9, Si—CH_3), 6.12 (doublet, 1, vinyl H, $J = 2$), 6.21 (doublet, 1, vinyl H, $J = 2$).

12. The checkers terminated the distillation when the head temperature reached 44°.

13. The oxidizing reagent is prepared as described by J. Meinwald, J. Crandall, and W. E. Hymans, *Org. Syn.*, Coll. Vol. **5,** 866 (1973).

14. Gas chromatographic analysis of the product by the submitters on a 1.85-m. 3% silicone gum rubber (SE-30) column at 25° gave a single peak. Butenone **4** has the following proton magnetic resonance spectrum (carbon tetrachloride) δ (multiplicity, number of protons, assignment, coupling constant J in Hz.): 0.14 (singlet, 9, Si—CH_3), 2.23 (singlet, 3, C—CH_3), 6.18 (doublet, 1, vinyl H, $J = 2$), 6.53 (doublet, 1, vinyl H, $J = 2$).

15. This material showed no tendency to deteriorate when stored under an argon atmosphere at −20°.

3. Discussion

Butenone **4** has been obtained by Brook and Duff[3] from the reaction of 1-trimethylsilylvinylmagnesium bromide and acetic anhydride at −120°. However, the product was a mixture of the butenone **4** and a dimeric substance which apparently resulted from subsequent conjugate addition of the Grignard reagent to the ketone. The procedures in the present reaction sequence for the preparation of silane **1**, silane **2**, and 1-trimethylsilylvinylmagnesium bromide are based on those reported by Ottolenghi, Fridkin, and Zilkha.[4] Other 1-trimethylsilylvinyl ketones may be similarly prepared by reaction of the appropriate aldehyde with 1-trimethylsilylvinylmagnesium bromide and subsequent oxidation with chromic acid. The 1-trimethylsilylvinyl ketones are remarkably stable and useful in a variety of conjugate addition reactions.[5]

1. Department of Chemistry, Wayne State University, Detroit, Michigan 48202.
2. Department of Chemistry, Cornell University, Ithaca, New York 14853.
3. A. G. Brook and J. M. Duff, *Can. J. Chem.*, **51**, 2024 (1973).
4. A. Ottolenghi, M. Fridkin, and A. Zilkha, *Can. J. Chem.*, **41**, 2977 (1963).
5. R. K. Boeckman, Jr., D. M. Blum, and B. Ganem, *Org. Syn.*, **58**, 158 (1978).

Appendix
Chemical Abstracts Nomenclature (Collective Index Number; Registry Numbers)

3-Buten-2-one, 3-(trimethylsilyl)- (8,9); (43209-86-5)

Silane, trimethylvinyl- (8); Silane, ethenyltrimethyl- (9); (754-05-2)

Vinylbromide: Ethylene, bromo- (8); Ethene, bromo- (9); (593-60-2)

Silane, chlorotrimethyl- (8,9); (75-77-4)

Silane, (1-bromovinyl)trimethyl- (8); Silane, (1-bromoethenyl)-trimethyl- (9); (13683-41-5)

Diethylamine (8); Ethanamine, *N*-ethyl- (9); (109-89-7)

3-Buten-2-ol, 3-(trimethylsilyl)- (8,9); (−)

Ethane, 1,2-dibromo- (8,9); (106-93-4)

Acetaldehyde (8,9); (75-07-0)

Chromic acid (H_2CrO_4) (8,9); (7738-94-5)

Sulfuric acid (8,9); (7664-93-9)

Isopropyl alcohol (8); 2-Propanol (9); (67-63-0)

Sodium ketyl of Benzophenone: Benzophenone, radical ion (1-), sodium (8); Methanone, diphenyl-, radical ion (1-), sodium (9); (3463-17-0)

1-Trimethylsilylvinylmagnesium bromide: Magnesium, bromo 1-(trimethylsilyl)vinyl (8); Magnesium, bromo[1-(trimethylsilyl)-ethenyl]- (9); (49750-22-3)

Acetic anhydride (8); Acetic acid, anhydride (9); (108-24-7)

3-TRIMETHYLSILYL-3-BUTEN-2-ONE AS MICHAEL ACCEPTOR FOR CONJUGATE ADDITION–ANNELATION: cis-4,4a,5,6,7,8-HEXAHYDRO-4a,5-DIMETHYL-2(3H)-NAPHTHALENONE

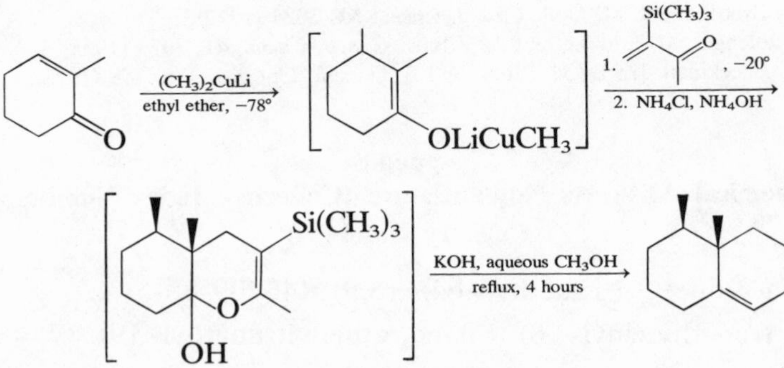

Submitted by ROBERT K. BOECKMAN, JR.,[1] DAVID M. BLUM,[1] and BRUCE GANEM[2]
Checked by SEIICHI INOUE and ROBERT M. COATES

1. Procedure

A 100-ml., three-necked flask is fitted with an argon inlet, a rubber septum, a magnetic stirrer, and a 25-ml. pressure-equalizing dropping funnel (Note 1). The flask is charged with 1.9 g. (0.010 mole) of cuprous iodide (Note 2) and 40 ml. of anhydrous ethyl ether. The mixture is stirred and cooled in an ice bath while 10 ml. (0.020 mole) of a $2 M$ solution of methyllithium in ethyl ether (Note 3) is injected through the septum into the flask. The resulting straw-yellow solution of lithium dimethylcuprate is cooled to $-78°$ in a dry ice–acetone bath, and a solution of 1.10 g. (0.010 mole) of 2-methyl-2-cyclohexenone (Note 4) in 10 ml. of ethyl ether is injected into the flask with stirring over a 2–3 minute period. The cooling bath is allowed to warm slowly to about $-20°$ over an interval of *ca.* 1 hour (Note 5), and then a solution of 2.13 g. (0.015 mole) of 3-trimethylsilyl-3-buten-2-one (Note 6) in 10 ml. of ethyl ether is added dropwise during 5 minutes. The mixture is stirred and cooled between $-20°$ and $-30°$ for another hour by occasional addition of dry ice to the cooling bath. The contents of the flask are poured into 100 ml. of an ammonium chloride–ammonium hydroxide buffer solution (Note 7) that has been cooled to 0° (Note 8). The ether layer is extracted with two or three additional 100-ml. portions of buffer solution (Note 9), and the combined aqueous solutions are extracted with two 75-ml. portions of ethyl ether. The combined ether extracts are washed once with saturated sodium chloride, dried with anhydrous magnesium sulfate, and evaporated at reduced pressure.

The residual yellow liquid (2.7–3.4 g.) is dissolved in a mixture of 40 ml. of methanol and 5 ml. of 4% aqueous potassium hydroxide. The resulting solution is heated at reflux under an argon atmosphere for 4 hours. The methanol is evaporated from the cooled solution under reduced pressure, and the residue is dissolved in 50 ml. of ethyl ether. The ether solution is washed once with water, dried with anhydrous magnesium sulfate, and evaporated. Distillation of the residual liquid with a Kugelrohr apparatus (Note 10) at 0.5 mm. affords, after separation of a 0.1–0.2 g. forerun collected at an oven temperature of 50°, 0.76–1.02 g. (43–57%) of the octalone at an oven temperature of 85–90° (Notes 11 and 12).

2. Notes

1. The apparatus is flamed dry and maintained under an atmosphere of argon during the reactions.

2. The submitters purified the cuprous iodide by precipitation from a concentrated aqueous solution of potassium iodide as described by Kauffman and Teter,[3] and dried it at 100° over phosphorous pentoxide at high vacuum. Cuprous iodide from Fisher Scientific Company was used by the checkers after drying at high vacuum.

3. A solution of methyllithium in ethyl ether may be purchased from Ventron Corporation. Directions for the preparation of ethereal methyllithium from methyl bromide are also available.[4] The solution should be standardized before use by a titration procedure such as that of Watson and Eastham.[5,6]

4. 2-Methyl-2-cyclohexenone is prepared by the method of E. W. Warnhoff, D. G. Martin, and W. S. Johnson, *Org. Syn.*, Coll. Vol. **4**, 162 (1963).

5. This warming operation was effected by the checkers by

removing the original cooling bath and replacing it with another one that had been cooled to −30°. The bath temperature was then allowed to rise to −20° over *ca.* 1 hour.

6. 3-Trimethylsilyl-3-buten-2-one was prepared by the method of R. K. Boeckman, Jr., D. M. Blum, B. Ganem, and N. Halvey, *Org. Syn.*, **58**, 152 (1978).

7. The buffer solution is prepared by adding enough concentrated ammonium hydroxide to 10% ammonium chloride to raise the pH to 8.

8. The checkers recommend that the mixture be stirred for 30 minutes at 0° to dissolve the pasty precipitate which forms and to facilitate the following extractions.

9. The extractions with buffer solution should be continued until the characteristic blue color of cupric ion is no longer visible in the aqueous layer.

10. Kugelrohr distillation ovens produced by Büchi Glasapparatefabrik are available from Brinkmann Instruments, Inc.

11. The checkers collected the 0.1–0.2 g. forerun at an oven temperature of about 70–85° and the main fraction over a 5–10° range between about 105 and 120°. Combination of the main fraction from two runs gave 2.03 g., which, on redistillation with the Kugelrohr apparatus at 0.4 mm., provided a 0.2 g. forerun collected with an oven temperature of 45–53° and a main fraction amounting to 1.5 g., which was collected at 88–93°.

12. The product has the following spectral properties: infrared (thin film) cm^{-1}: 2960 (C—H), 1685 (C=O), 1620 (C=C); proton magnetic resonance (carbon tetrachloride) δ (multiplicity, number of protons, assignment): 5.64 (broad singlet, 1, vinyl-H), 1.11 (singlet, 3, CH_3), 0.92 (multiplet, 3, CH_3), minor absorption of an unidentified impurity at 5.75 (multiplet, 0.1–0.2, vinyl-H).

A gas chromatographic analysis of the product by the submitters using a 1.8-m. column packed with 20% Carbowax 20 M suspended on Chromosorb P and operated at 150° with a flow rate of 30 ml. per minute showed a peak for the major component having a retention time of 16 minutes and two minor peaks having retention times of 4 and 7 minutes, with relative areas amounting to 6% and 2% of the major peak, respectively. The stereochemical purity of the product was shown to be >95% *cis* by the submitters by gas chromatographic analysis using a 50-ft. capillary column

TABLE I
Synthesis of Polycyclic Ketones[10–13]

Reactant	Product	Yield (%)
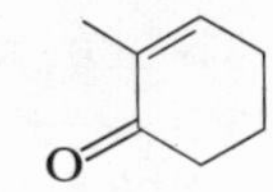	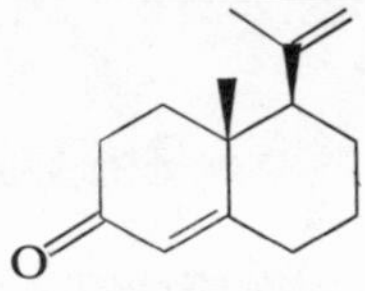	57
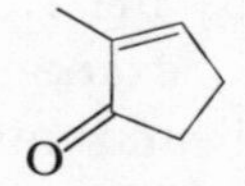	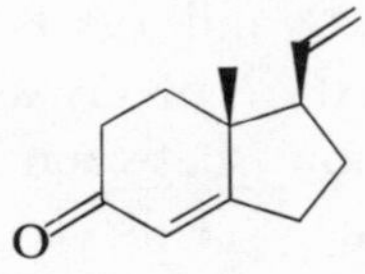	54
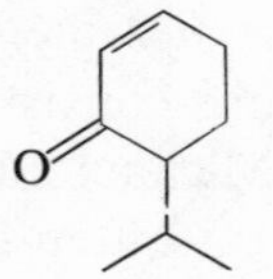	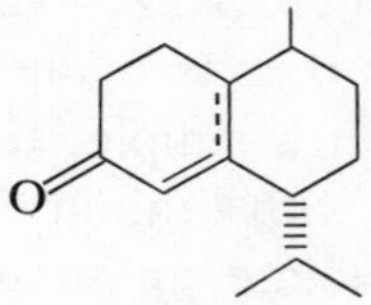	53
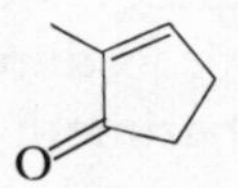	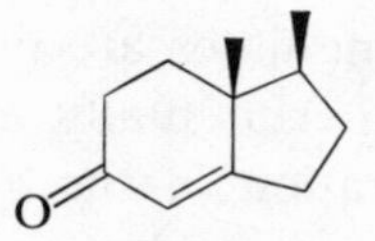	57
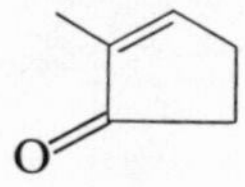	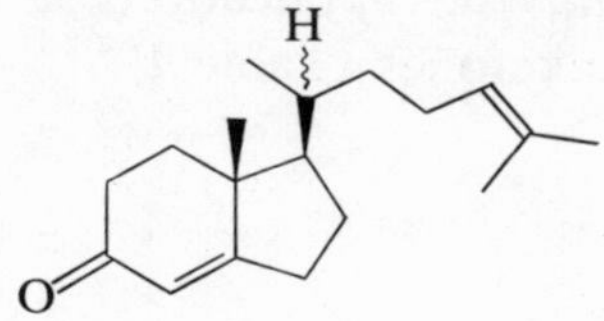	67
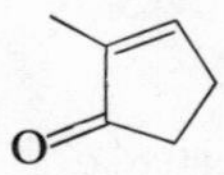	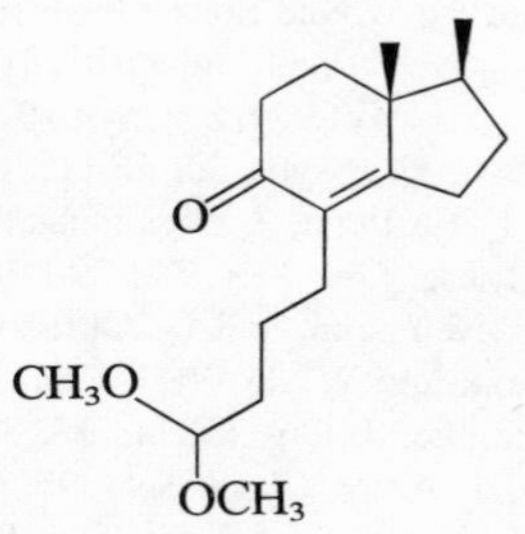	>60

coated with Carbowax 20 M and heated to 140°; these conditions give separate peaks for a 3:2 mixture of the *cis* and *trans* isomers of the two octalones.[7,8]

3. Discussion

4,4*a*,5,6,7,8-Hexahydro-4*a*,5-dimethyl-2(3*H*)-naphthalenone, an important intermediate in the total syntheses of the sesquiterpenes, aristolone,[7,8] and fukinone,[9] has been prepared in 15% yield as a 3:2 mixture of *cis* and *trans* isomers by the Robinson annelation reaction between 2,3-dimethylcyclohexanone and methyl vinyl ketone.[7,8] This *cis*-octalone has also been synthesized stereoselectively via the crystalline enol lactone, *cis*-3,4,4*a*,5,6,7-hexahydro-4*a*,5-dimethyl-2*H*-1-benzopyran-2-one.[8] In the procedure reported here the *cis*-octalone is prepared from 2-methyl-2-cyclohexen-1-one in a simple, three-step procedure consisting of conjugate addition with lithium dimethylcuprate, Michael addition of the resulting enolate anion to 3-trimethylsilyl-3-buten-2-one, and cyclization of the intermediate diketone.[10] Reasonable structures for the intermediates are proposed in the lead equation but have not been experimentally established. This procedure also serves as an illustration of the general utility of 1-trimethylsilylvinyl ketones in regio- and stereoselective enolate trapping reactions[11] and the application of the method in the synthesis of polycyclic ketones (Table I).[10-13]

1. Department of Chemistry, Wayne State University, Detroit, Michigan, 48202.
2. Department of Chemistry, Cornell University, Ithaca, New York 14853
3. G. B. Kauffman and L. A. Teter, *Inorg. Syn.*, **7**, 9 (1963).
4. G. Wittig and A. Hesse, *Org. Syn.*, **50**, 66 (1970).
5. S. C. Watson and J. F. Eastham, *J. Organometal. Chem.*, **9**, 165 (1967).
6. M. Gall and H. O. House, *Org. Syn.*, **52**, 39 (1972).
7. C. Berger, M. Franck-Neumann, and G. Ourisson, *Tetrahedron Lett.*, 3451 (1968).
8. E. Piers, R. W. Britton, and W. de Waal, *Can. J. Chem.*, **47**, 4307 (1969).
9. E. Piers and R. D. Smillie, *J. Org. Chem.*, **35**, 3997 (1970).
10. R. K. Boeckman, Jr., *J. Amer. Chem. Soc.*, **95**, 6867 (1973).
11. G. Stork and B. Ganem, *J. Amer. Chem. Soc.*, **95**, 6152 (1973).
12. R. K. Boeckman, Jr., *J. Amer. Chem. Soc.*, **96**, 6179 (1974).
13. G. Stork and J. Singh, *J. Amer. Chem. Soc.*, **96**, 6181 (1974).

Appendix
Chemical Abstracts Nomenclature (Collective Index Number; Registry Numbers)

3-Buten-2-one, 3-trimethylsilyl- (8,9); (43209-86-5)

2(3H)-Naphthalenone, 4,4a,5,6,7,8-hexahydro-4a,5-dimethyl-, cis, ($\pm$)- (8,9); (20536-80-5)

Lithium, methyl- (8,9); (917-54-4)

Cuprate (1-), dimethyl-, lithium (8,9); (15681-48-8)

2-Cyclohexen-1-one, 2-methyl- (8,9); (1121-18-2)

Aristolone: 2H-Cyclopropa[a]naphthalen-2-one, 1,1$a\beta$,4,5,6,7,7a, 7$b\beta$-octahydro-1,1,7β,7$a\beta$-tetramethyl- (8); 2H-Cyclopropa[a]-naphthalen-2-one, 1,1a,4,5,6,7,7a,7b-octahydro-1,1,7,7a-tetra-methyl-, (1$a\alpha$,7α,7$a\alpha$,7$b\alpha$) (9); (6831-17-0)

Fukinone: Eremophil-7(11)-en-8-one (8); 2(1H)-Naphthalenone, octahydro-4a,5-dimethyl-3-(1-methylethylidene)-,[4aR-(4a,5,8a)]- (9); (19593-06-7)

Cyclohexanone, 2,3-dimethyl- (8,9); (13395-76-1)

Methyl vinyl ketone: 3-Buten-2-one (8,9); (78-94-4)

2H-1-Benzopyran-2-one, 3,4,4a,5,6,7-hexahydro-4a,5-dimethyl-, cis- (8,9); (51557-48-3)

2-TRIMETHYLSILYLOXY-1,3-BUTADIENE AS A REACTIVE DIENE: DIETHYL $trans$-4-TRIMETHYLSILYLOXY-4-CYCLOHEXENE-1,2-DICARBOXYLATE

(4-Cyclohexene-1,2-dicarboxylic acid, 4-trimethylsilyloxy-, diethyl ester)

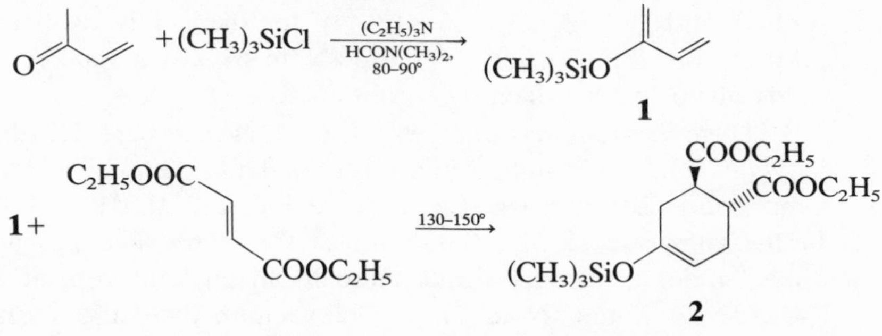

Submitted by MICHAEL E. JUNG and CHARLES A. MCCOMBS[1]
Checked by PING SUN CHU and GEORGE H. BÜCHI

1. Procedure

*Caution! Part A should be carried out in a hood, since the
reagents are noxious.*

A. *2-Trimethylsilyloxy-1,3-butadiene* (**1**).   An oven-dried 500-
ml., three-necked, round-bottomed flask is fitted with two oven-
dried addition funnels, a glass stopper, and magnetic stirrer, and
placed in a 80–90° oil bath. Under an inert atmosphere, methyl
vinyl ketone (25.0 g., 0.357 mole) in 25 ml. of dimethylformamide
and chlorotrimethylsilane (43.4 g., 0.400 mole) in 25 ml. of
dimethylformamide are added over 30 minutes to a magnetically
stirred solution of triethylamine (40.5 g., 0.400 mole) in 200 ml. of
dimethylformamide (Note 1). The reaction gradually darkens from
colorless to yellow or dark brown, and supports a white precipitate
of triethylamine hydrochloride. The reaction is set up to run
overnight, or *ca.* 14 hours.

The reaction is cooled to room temperature, filtered (Note 2),
and transferred to a 2-l. separatory funnel containing 300 ml. of
pentane. To this solution is added 1 l. of *cold* 5% sodium bicarbonate solution to facilitate the separation of phases and remove the
dimethylformamide. The pentane layer is separated and the aqueous layer extracted twice with 300-ml. portions of pentane. The
pentane extracts are combined, washed with 200 ml. of *cold* distilled water (Note 3), dried over powdered anhydrous sodium sulfate,
and filtered into a 2-l. round-bottomed flask.

The pentane and other volatiles are removed by fractional
distillation using a 5-cm. steel-wool-packed column and heating
the pot in a 70° oil bath (Note 4). A water aspirator vacuum is
applied, and 18.2–22.9 g. (36–45%) of the diene **1** is distilled as a
colorless oil, b.p. 50–55° (50 mm.) (Note 5). On a smaller scale,
yields of up to 50% have been obtained.

B. *Diethyl* trans-4-*trimethylsilyloxy*-4-*cyclohexene*-1,2-*dicar-
boxylate* (**2**).   A 25-ml., round-bottomed flask equipped with a
reflux condenser is charged with diene **1** (7.1 g., 0.050 mole) and
diethyl fumarate (5.7 g., 0.033 mole) (Note 6). The mixture is
stirred under a nitrogen atmosphere in an oil bath kept at 130–
150° for 24 hours (Note 7). Direct vacuum distillation using a
short-path distillation apparatus affords a small amount of lower-

boiling material and then 8.02 g. (77%) of cyclohexene **2**, b.p. 127–128° (0.5 mm.) (Notes 8 and 9).

2. Notes

1. The checkers obtained methyl vinyl ketone from Hoffmann-La Roche Company and chlcrotrimethylsilane from Aldrich Chemical Company, Inc. Both were distilled from calcium hydride. Triethylamine obtained from J. T. Baker Chemical Company was distilled from calcium hydride, and dimethylformamide from Fisher Scientific Company was used from a freshly opened bottle.

2. Filtration was conveniently performed through a plug of glass wool directly into the separatory funnel.

3. The bicarbonate extractions must be performed quickly, since the product slowly hydrolyzes in the presence of water. The best yields were obtained when the phases were shaken briskly for 10 seconds and separated as soon as foaming ceased. Foaming also occurred when the pentane extracts were washed with water, but did not prevent the separation of phases.

4. The checkers used a 18-cm. Vigreux column to remove low-boiling material, and then transferred the residue to a 250-ml. round-bottomed flask for fractional distillation under reduced pressure.

5. The distilled diene **1** has been obtained with 99% purity by the submitters. The product has the following spectral properties: proton magnetic resonance: (carbon tetrachloride with benzene as internal standard) δ (multiplicity, number of protons): 0.4 (singlet, 9), 4.4 (broad singlet, 2), 5.1 (doublet of multiplets, 1), 5.5 (doublet of doublets, 1), 6.3 (doublet of doublets, 1). The impurities were triethylamine and hexamethyldisiloxane. When stored in a serum-capped flask and removed via syringe, the butadiene is stable for 2 months in a desiccator.

6. The checkers prepared diethyl fumarate by the sulfuric acid–catalyzed esterification of fumaric acid.[7] Diethyl fumarate obtained from Aldrich Chemical Company, Inc., contained dimethyl fumarate as a contaminant.

7. The submitters allowed the reaction to proceed for 24 hours after monitoring by gas chromatography (1.83 m. × 0.64 cm., 10%

SE-30/60–80 mesh Chromosorb W, 150°) indicated that the disappearance of diethyl fumarate was not complete after 5.5 hours. The submitters made no attempt to optimize the reaction time. The checkers found the reaction to be complete after 7 hours at 135–145°.

8. The checkers distilled cyclohexene **2** after 7 hours' reaction time using a Büchi Kugelrohr apparatus to obtain 9.9–10.0 g. (95–96%) of cyclohexene **2**.

9. Cyclohexene **2** has the following spectral properties: infrared (liquid film) cm^{-1}: 1730, 1670; proton magnetic resonance: (carbon tetrachloride) δ (multiplicity, number of protons, assignments, coupling constant J in Hz.): 0.16 [singlet, 9, Si(CH$_3$)$_3$], 1.17 (triplet, 6, CH$_3$, $J = 7$), 2.00–3.16 (multiplet, 6, ring CH), 4.07 (quartet, 4, OCH$_2$, $J = 7$), 4.70 (broad singlet, 1, olefinic CH).

3. Discussion

The first reference to 2-trimethylsilyloxy-1,3-butadiene (**1**) was a report[2] of its reaction with tetracyanoethylene by Cazeau and Frainnet without mention of any experimental details. Later, Conia[3] reported its synthesis in 50% yield with only a reference made to the usual House procedure[4] for silyl enol ethers. The diene **1** has also been prepared using lithium diisopropylamide as base and chlorotrimethylsilane in tetrahydrofuran–ether (1:1) in yields up to 65%, but on a smaller scale.[5]

TABLE I
PREPARATION OF TRIMETHYLSILYLOXYCYCLOHEXENES

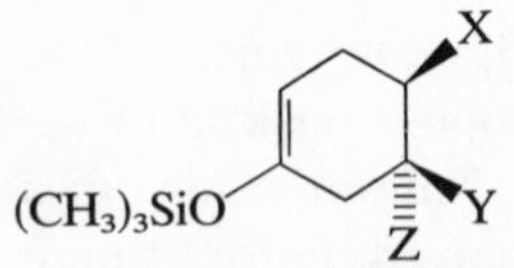

X	Y	Z	Yield (%)
COCH$_3$	H	H	60
CO$_2$CH$_3$	H	H	46
CO$_2$CH$_3$	H	CO$_2$CH$_3$	71
CO$_2$C$_2$H$_5$	H	CO$_2$C$_2$H$_5$	77
CO$_2$C$_2$H$_5$	CO$_2$C$_2$H$_5$	H	39

Butadienes substituted with alkoxy groups in the 2-position, *e.g.*, 2-ethoxy-1,3-butadiene,[6] have been prepared from methyl vinyl ketone, but they required several conversions and a tedious spinning-band distillation to purify the product. This slight modification of the House procedure has been used to conveniently prepare 2-trimethylsilyloxy-1,3-butadiene from the readily available methyl vinyl ketone. This one-step procedure has provided large amounts of a new and reactive diene for Diels–Alder reactions, as illustrated in Table I.

1. Contribution No. 3759 from the Department of Chemistry, University of California, Los Angeles, California 90024.
2. P. Cazeau and E. Frainnet, *Bull. Soc. Chim. Fr.*, 1658 (1972).
3. C. Girard, P. Amice, J. P. Barnier, and J. M. Conia, *Tetrahedron Lett.*, 3329 (1974).
4. H. O. House, L. J. Czuba, M. Gall, and H. D. Olmstead, *J. Org. Chem.*, **34**, 2324 (1969).
5. M. E. Jung and C. A. McCombs, *Tetrahedron Lett.*, 2935 (1976).
6. (a) H. B. Dykstra, *J. Amer. Chem. Soc.*, **57**, 2255 (1935); (b) N. A. Milas, E. Sakel, J. T. Plati, J. T. Rivers, J. K. Gladding, F. X. Grossi, Z. Weiss, M. A. Campbell, and H. F. Wright, *J. Amer. Chem. Soc.*, **70**, 1597 (1948).

Appendix
Chemical Abstracts Nomenclature (Collective Index Number; Registry Numbers)

2-Trimethylsilyloxy-1,3-butadiene: Silane, trimethyl[(1-methylene-2-propenyl)oxy]- (8,9); (38053-91-7)

Methyl vinyl ketone: 3-Buten-2-one (8,9); (78-94-4)

Formamide, *N,N*-dimethyl- (8,9); (68-12-2)

Silane, chlorotrimethyl- (8,9); (75-77-4)

Triethylamine (8); Ethanamine, *N,N*-diethyl- (9); (121-44-8)

Diethyl *trans*-4-trimethylsilyloxy-4-cyclohexene-1,2-dicarboxylate: 4-Cyclohexene-1,2-dicarboxylic acid, 4-(trimethylsilyloxy)-, diethyl ester (8,9); (−)

Diethyl fumarate: Fumaric acid, diethyl ester (8); 2-Butenedioic acid (*E*)-, diethyl ester (9); (623-91-6)

Disiloxane, hexamethyl- (8,9); (107-46-0)

Fumaric acid (8); 2-Butenedioic acid (*E*)-(9); (110-17-8)

Dimethyl fumarate: Fumaric acid, dimethyl ester (8); 2-Butenedioic acid (*E*)- dimethyl ester (9); (624-49-7)

Tetracyanoethylene: Ethenetetracarbonitrile (8,9); (670-54-2)

Lithium diisopropylamide: Diisopropylamine, lithium salt (8); 2-propanamine, N-(1-methylethyl)-, lithium salt (9); (4111-54-0)

1,3-Butadiene, 2-ethoxy- (8,9); (4747-05-1)

CHEMICAL HAZARD WARNING

Benzene has been identified as a carcinogen. OSHA has issued emergency standards (5/77) on its use. All procedures involving benzene should be carried out in a well-ventilated hood, and glove protection is required.

For an OSHA list of carcinogens see *Organic Syntheses*, **56,** 128 (1977). The following compounds have been added to the OSHA list of carcinogens:

   *Asbestos
   Benzene
   *Coal tar volatiles—coke oven emissions
   Vinyl chloride

The following have also been identified as strong carcinogens:

   Benz(a)pyrene
   2,4-Diaminotoluene
   Dimethylcarbamoyl chloride
   1,1-Dimethylhydrazine (and salts)
   Dimethyl sulfate
   Hexamethylphosphoramide
   Hydrazine (and salts)
   N-(2-Hydroxyethyl)ethyleneimine
   Methylhydrazine (and salts)
   2-Nitronaphthalene
   Nitrosoamines
   Propane sultone
   Propyleneimine

* Standards classifying these as carcinogens have been proposed, but have not been adopted.

The following compounds have been identified as experimental carcinogens:

Acrylonitrile
3-Amino-1,2,4-triazole
Carbon tetrachloride
Chloroform
1,4-Dichloro-2-butene
Dioxane
Epichlorohydrin
Ethylene dibromide
Ethylenethiourea
Lead chromate
Methylenedianiline
Styrene
Tetramethyl thiourea
Thiourea
o-Toluidene
Trichloroethylene
Vinylcyclohexene dioxide
Zinc chromate

CUMULATIVE AUTHOR INDEX
FOR VOLUMES 50 TO 58

This index comprises the names of contributors to Volumes 50 through 58. A number in **boldface type** denotes the volume; a number in ordinary type indicates the page of that volume. For authors to previous volumes see *Organic Syntheses*, cumulative indices for Collective Volumes 1-5 [Ed. R. L. Shriner and R. H. Shriner (1976)] and cumulative index in volume 54 (volumes 50-54).

CUMULATIVE SUBJECT INDEX
FOR VOLUMES 55 TO 58

This index comprises material from Volumes 55 through 58; for subjects of previous volumes see cumulative indices for Collective Volumes 1–5 and the cumulative index in Volume 54 (Volumes 50–54).

This Index consists of two parts.

Part I contains entries referring to the names of compounds as used in these volumes followed by page number.

Part II contains entries referring to the names of compounds according to Chemical Abstracts Systematic Nomenclature (see Index Guide, Chemical Abstracts Vol. 76, 1972) followed by registry number and page number.

Entries in capital letters in Part I and Part II indicate compounds in the title of a preparation. Entries in ordinary type letters refer to principal products and major by-products, special reagents or intermediates (which may or may not be isolated), compounds mentioned in the text, Notes or Discussion as having been prepared by the method given, and apparatus described in detail or illustrated by a figure. Numbers in boldface type denote the volumes. Numbers in ordinary type indicate pages on which a compound or subject is mentioned in the indicated volume. Compounds in the Tables given by formula only may not be indexed.

PART I

(Nomenclature Used in These Volumes)

Abietylamine, dehydro- N-trifluoro-, **56**, 125

Acenaphthylene, **58**, 73

Acetaldehyde, **58**, 157

Acetamides, N-arylalkyl-, **56**, 7

Acetanilide, 2,2,2-trifluoro-, **56**, 122

Acetic acid, butyryl-, ethyl ester, **55**, 73, 75

Acetic acid, chloro-, *tert*-butyl ester, **55**, 94

Acetic acid, cyano-, ethyl ester, **55**, 58, 60

Acetic acid, cyano-, methyl ester, **56**, 63

Acetic acid, 3,4-dimethoxy-phenyl-, **55**, 45, 46

Acetic acid, methoxy-, **56**, 70

Acetic acid, nitro-, dipotassium salt, **55**, 77, 78

ACETIC ACID, NITRO-, METHYL ESTER, **55**, 77, 78

Acetic acid, phenoxy-, **56**, 68

Acetic acid, phenyl-, **56**, 70

Acetic acid, phenyl ester, **56**, 126

Acetic acid, trifluoro-, **55**, 70

Acetic acid, trifluoro-, anhydride, **56**, 125

Acetic acid, vinyl-, **56**, 49

Acetic anhydride, **58**, 157

Acetone, **58**, 138

Acetone, amino-, semicarbazone, hydrochloride, correction note, **56**, 127

PART II

(Chemical Abstracts Systematic Nomenclature)

Unchecked Procedures

Received during the period July 1, 1977–June 30, 1978
and subsequently accepted for checking*

In accordance with a policy adopted by the Board
of Editors, beginning with Volume 50 and further modified as noted in the Editor's Preface in Volume 55,
procedures received by the Secretary during the year
and subsequently accepted for checking by Organic
Syntheses, will be made available for purchase at the
price of $2 per procedure, prepaid, upon request to
the Secretary:

> Dr. Wayland E. Noland, Secretary
> Organic Syntheses
> University of Minnesota
> Department of Chemistry
> 207 Pleasant Street S. E.
> Minneapolis, Minnesota 55455

Payment must accompany the order, and should be
made payable to Organic Syntheses, Inc. (not to the
Secretary). Purchase orders not accompanied by payment will not be accepted. Procedures may be ordered
by number and/or title from the list which follows.

It should be emphasized that the procedures
which are being made available are unedited and have
been reproduced just as they are first received from
the submitters. There is no assurance that the procedures listed here will ultimately check in the form
available, and some of them may be rejected for publication in Organic Syntheses during or after the
checking process. For this reason, Organic Syntheses
can provide no assurance whatsoever that the procedures will work as described, and offers no comment
as to what safety hazards may be involved. Consequently, more than usual caution should be employed
in following the directions in the procedures.

Organic Syntheses welcomes, on a strictly volun-
tary basis, comments from persons who attempt to carry
out the procedures. For this purpose, a Checker's
Report form will be mailed out with each unchecked
procedure ordered. Procedures which have been
checked by or under the supervision of a member of
the Board of Editors will continue to be published
in the volumes of Organic Syntheses, as in the past.
It is anticipated that many of the procedures in the
list will be published (often in revised form) in
Organic Syntheses in future volumes.

*   A procedure marked with a superscript a has been checked.
Procedures received before June 30, 1977, but accepted between July 1, 1977 and June 30, 1978 are marked with a superscript b.

2035[a,b]   Removal of N^{α}-Benzyloxycarbonyl Groups from Sulfur-Containing Peptides by Catalytic Hydrogenation in Liquid Ammonia

A. M. Felix, M. H. Jimenez, and J. Meienhofer, Research Division, Hoffmann-La Roche Inc., Nutley, NJ 07110

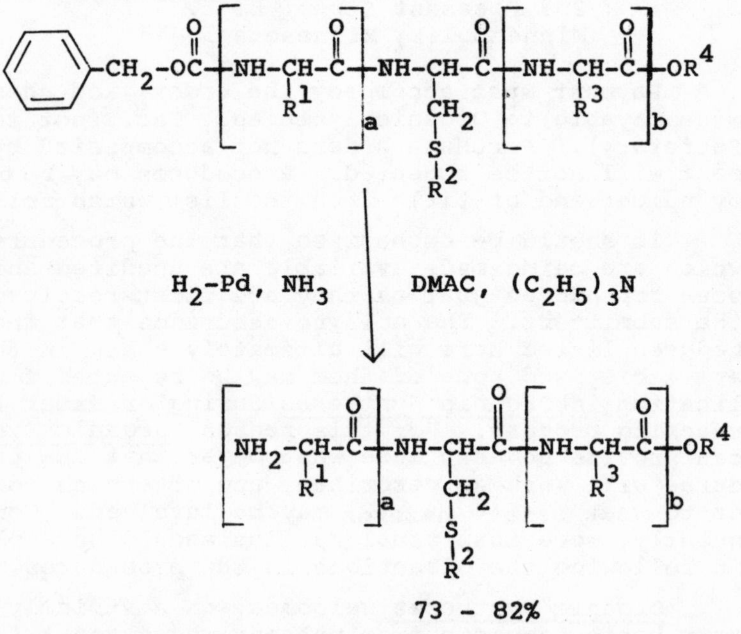

H_2-Pd, NH_3 DMAC, $(C_2H_5)_3N$

73 - 82%

R^1, R^3 represent the natural amino acids with tert-butyl derived protecting groups

R^2 = Bzl, Acm, tert-Bu, CH_3

R^4 = H, tert-Bu

2036[b] Bis(2,2,2-trichloroethyl) Azodicarboxylate

R. D. Little and M. G. Venegas, Department of Chemistry, University of California, Santa Barbara, Santa Barbara, CA 93106

$$H_2NNH_2 \cdot H_2O + 2ClCO_2CH_2CCl_3 + Na_2CO_3 \longrightarrow$$

$$Cl_3CCH_2O_2CNHNHCO_2CH_2CCl_3 + 2NaCl + CO_2 + 2H_2O$$

93%

$$Cl_3CCH_2O_2CNHNHCO_2CH_2CCl_3 \xrightarrow[\text{HNO}_3 \ (\text{fuming})]{[O]}$$

$$Cl_3CCH_2O_2CN=NCO_2CH_2CCl_3$$

76%

2040[b] Cyclobutanone

M. Krumpolc and J. Rocek, Department of Chemistry, University of Illinois at Chicago Circle, Box 4348, Chicago, IL 60680

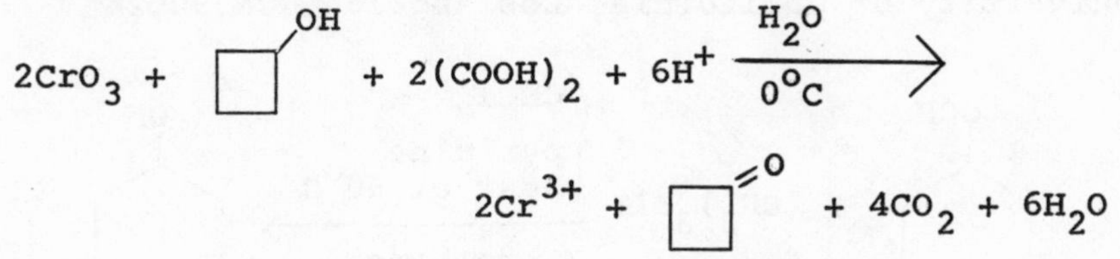

$$2CrO_3 + \underset{}{\square}\text{-OH} + 2(COOH)_2 + 6H^+ \xrightarrow[0°C]{H_2O}$$

$$2Cr^{3+} + \underset{}{\square}{=}O + 4CO_2 + 6H_2O$$

85-92%

2043[a,b] Iodotrimethylsilane

M. E. Jung and M. A. Lyster, Department of Chemistry,
University of California, Los Angeles, CA 90024

A. $(CH_3)_3SiCl + C_6H_5N(CH_3)_2 \xrightarrow[\substack{\text{bath} \\ \text{temperature,} \\ 125^\circ C}]{H_2O} [(CH_3)_3Si]_2O$ 93%

B. $[(CH_3)_3Si]_2 + Al + I_2 \xrightarrow[\substack{\text{bath} \\ \text{temperature} \\ 140^\circ C}]{} (CH_3)_3SiI$ 87.5%

2044[a,b] The Cleavage of Methyl Ethers Using Iodotrimethylsilane: Cyclohexanol from Cyclohexyl Methyl Ether

M. E. Jung and M. A. Lyster, Department of Chemistry,
University of California, Los Angeles, CA 90024

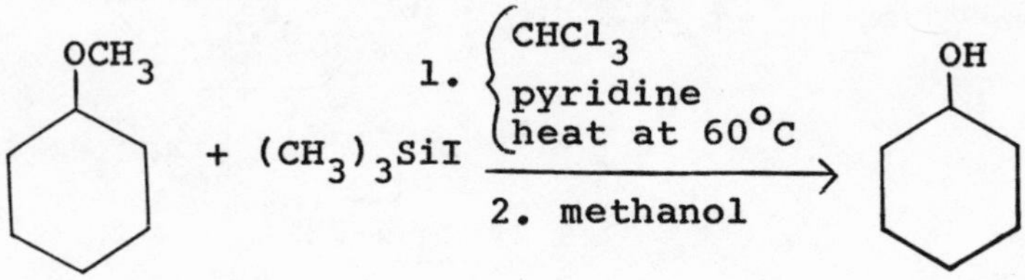

78-88%

2-Ethoxypyrrolin-5-one: <u>Tert-Butyl</u>
<u>N-(1-Ethoxycyclopropyl)carbamate</u>

G. C. Crockett and T. H. Koch, Department of Chemistry, University of Colorado, Boulder, CO 80309

A.

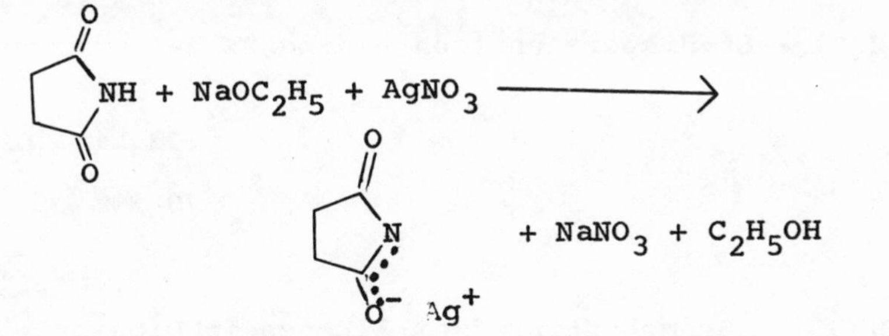

90-95%

B.

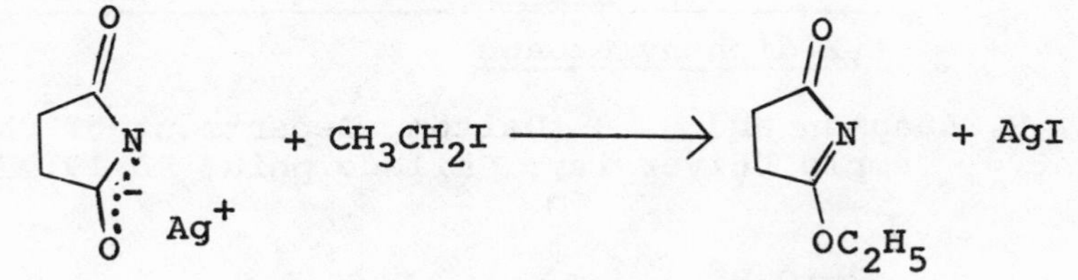

44-60%

C.

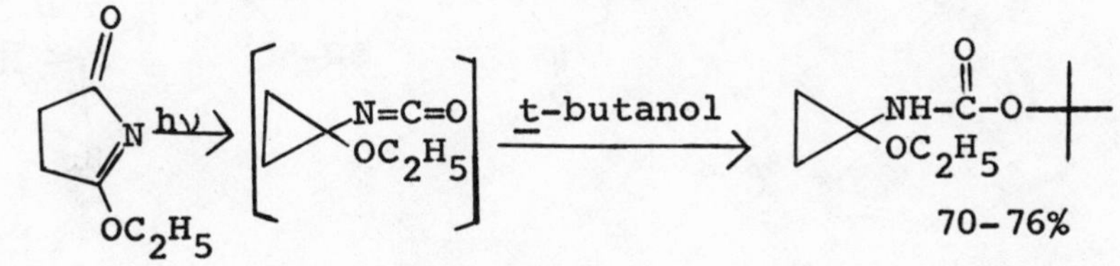

70-76%

2052 <u>1,6-Dimethyltricyclo[4.1.0.0²,⁷]hept-3-ene</u>

R. T. Taylor and L. A. Paquette, Department of Chemistry, The Ohio State University, Columbus, OH 43210

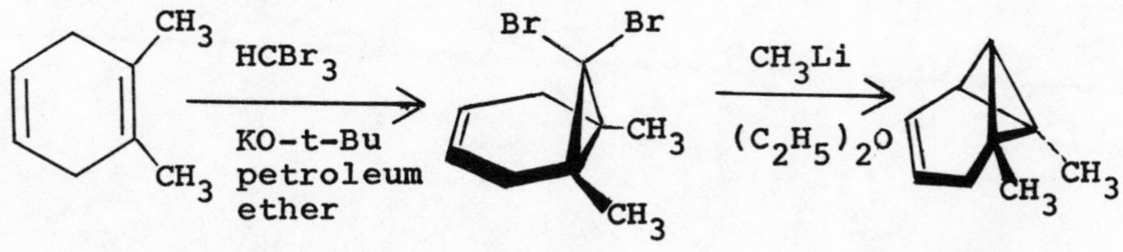

57-64%   46-55%

2054 Preparation of Halide-Free Methyllithium

M. J. Lusch, W. V. Phillips, R. F. Sieloff, and H. O. House, Department of Chemistry, Georgia Institute of Technology, Atlanta, GA 30332

$$CH_3Cl + \text{Li dispersion (containing 2\% Na)} \xrightarrow[25^{\circ}]{(C_2H_5)_2O} CH_3Li + LiCl$$

70–89%

2059[a] Bromohydrins from Alkenes in Dimethyl Sulfoxide: *erythro*-1-Bromo-2-hydroxy-1,2-diphenylethane

A. W. Langman and D. R. Dalton, Department of Chemistry, Temple University, Philadelphia, PA 19122

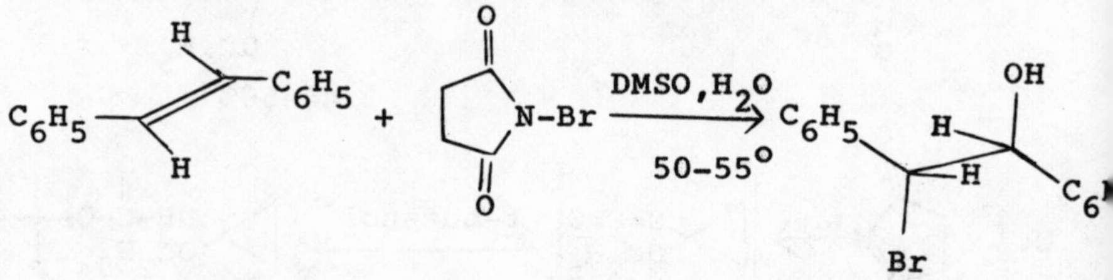

<u>Cyclopropylmalonaldehyde</u>

W. Pressler and C. Reichardt, Fachbereich Chemie, der Philipps-Universität, D-3550 Marburg 1, Lahnberge, Postfach 1929, Germany

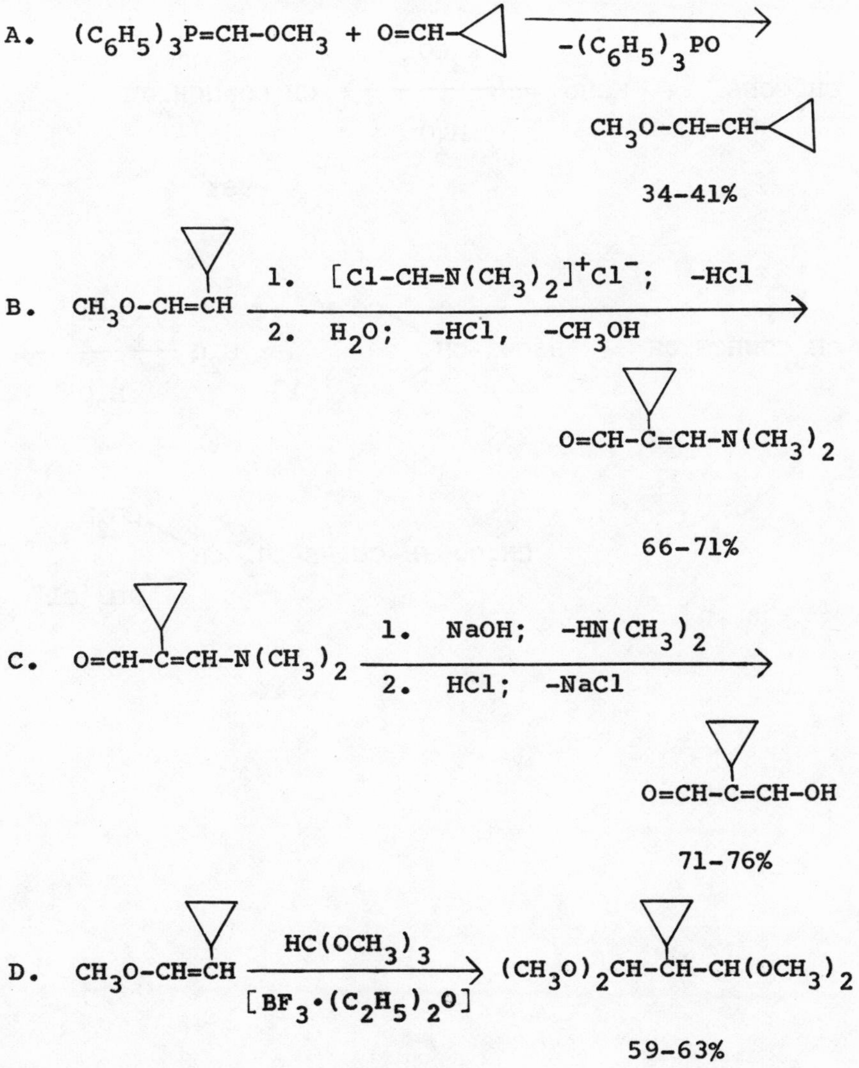

A. $(C_6H_5)_3P=CH-OCH_3$ + O=CH-◁ $\xrightarrow[-(C_6H_5)_3PO]{}$

$CH_3O-CH=CH-◁$

34-41%

B. $CH_3O-CH=CH-\overset{\displaystyle \triangledown}{|}$ $\xrightarrow[\text{2. } H_2O; \quad -HCl, \quad -CH_3OH]{\text{1. } [Cl-CH=N(CH_3)_2]^+Cl^-; \quad -HCl}$

$O=CH-\overset{\displaystyle \triangledown}{C}=CH-N(CH_3)_2$

66-71%

C. $O=CH-\overset{\displaystyle \triangledown}{C}=CH-N(CH_3)_2$ $\xrightarrow[\text{2. } HCl; \quad -NaCl]{\text{1. } NaOH; \quad -HN(CH_3)_2}$

$O=CH-\overset{\displaystyle \triangledown}{C}=CH-OH$

71-76%

D. $CH_3O-CH=CH-\overset{\displaystyle \triangledown}{|}$ $\xrightarrow[{[BF_3 \cdot (C_2H_5)_2O]}]{HC(OCH_3)_3}$ $(CH_3O)_2CH-\overset{\displaystyle \triangledown}{CH}-CH(OCH_3)_2$

59-63%

$\underline{S}$-Acetamidomethyl-L-cysteine

J. D. Milkowski, Merck Sharp & Dohme Research
Laboratories, P. O. Box 2000, Rahway, NJ 07065,
D. Veber and R. Hirschmann, Merck Sharp & Dohme
Research Laboratories, West Point, PA 19486

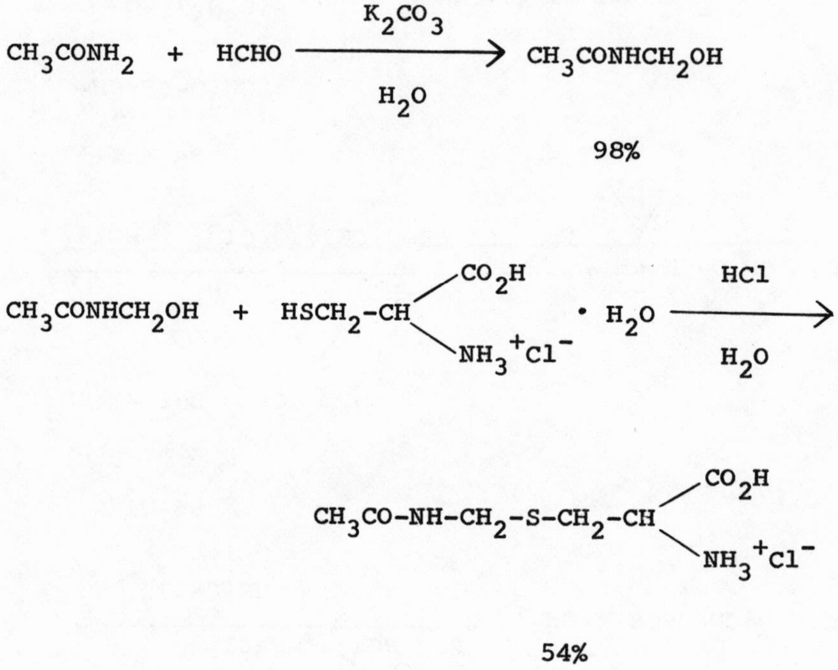

$$CH_3CONH_2 \ + \ HCHO \ \xrightarrow[\text{H}_2\text{O}]{\text{K}_2\text{CO}_3} \ CH_3CONHCH_2OH$$

98%

$$CH_3CONHCH_2OH \ + \ HSCH_2-CH{\overset{\displaystyle CO_2H}{\underset{\displaystyle NH_3^+Cl^-}{}}} \cdot H_2O \ \xrightarrow[\text{H}_2\text{O}]{\text{HCl}}$$

$$CH_3CO-NH-CH_2-S-CH_2-CH{\overset{\displaystyle CO_2H}{\underset{\displaystyle NH_3^+Cl^-}{}}}$$

54%

Alkynes from Organic Halides by Ion Pair Extraction Dehydrohalogenation: Propargyl Aldehyde Diethyl Acetal

A. Le Coq and A. Gorgues, Laboratoire de Synthese Organique, Universite de Rennes, Avenue du Général Leclerc, B. P. 25 A-35031, Rennes Cédex, France

$$(C_4H_9)_4N^+HSO_4^- \ + \ 2NaOH \ \Longleftarrow \Longrightarrow \ (C_4H_9)_4N^+OH^- \ + $$

$$Na_2SO_4 \ + \ H_2O$$

$$CH_2=CHCH{\overset{O}{\overset{\|}{}}} \ + \ Br_2 \ \xrightarrow[\ C_2H_5OH\ ]{\ C_2H_5O{\overset{O}{\overset{\|}{}}}CH\ }$$

$$CH_2Br-CHBr-CH(OC_2H_5)_2$$

78-84%

$$2(C_4H_9)_4N^+OH^- \ + \ CH_2Br-CHBr-CH(OC_2H_5)_2 \ \longrightarrow$$

$$H-C{\equiv}C-CH(OC_2H_5)_2 \ + \ 2(C_4H_9)_4N^+Br^- \ + \ 2H_2O$$

67-74%

R. C. Cambie and P. S. Rutledge, The University of
Auckland, Auckland, New Zealand

A.

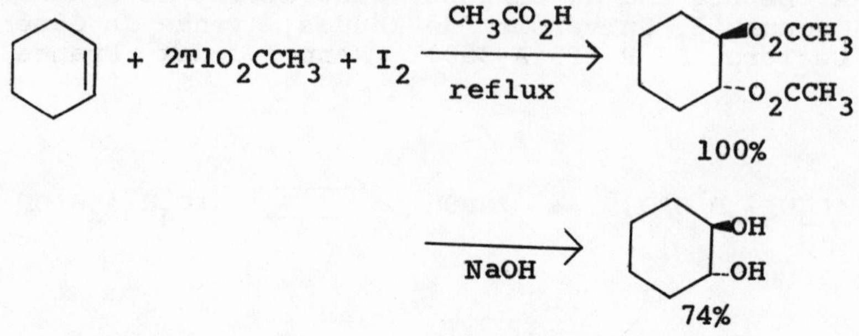

$$\text{cyclohexene} + 2\text{TlO}_2\text{CCH}_3 + \text{I}_2 \xrightarrow[\text{reflux}]{\text{CH}_3\text{CO}_2\text{H}}$$

trans-1,2-bis(acetoxy)cyclohexane (O_2CCH_3, O_2CCH_3)

100%

$$\xrightarrow{\text{NaOH}}$$ trans-cyclohexane-1,2-diol (OH, OH)

74%

B.

$$\text{cyclohexene} + 2\text{TlO}_2\text{CCH}_3 + \text{I}_2 \xrightarrow[\text{H}_2\text{O}]{\text{CH}_3\text{CO}_2\text{H}}$$

cis-product (O_2CCH_3, OH)

$$\xrightarrow{\text{NaOH}}$$ cis-cyclohexane-1,2-diol (OH, OH)

71%

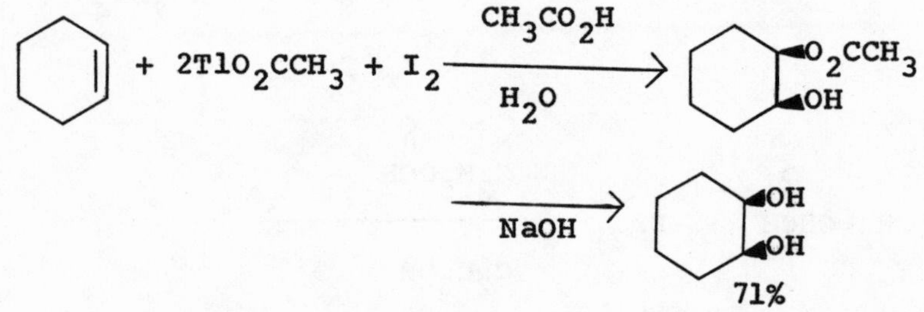

2069   Generation and Reactions of Vinyllithium
        Reagents: 2-Butylbornene

A. R. Chamberlin, E. L. Liotta, and F. T. Bond,
Department of Chemistry, University of California,
San Diego, D-006, La Jolla, CA 92093

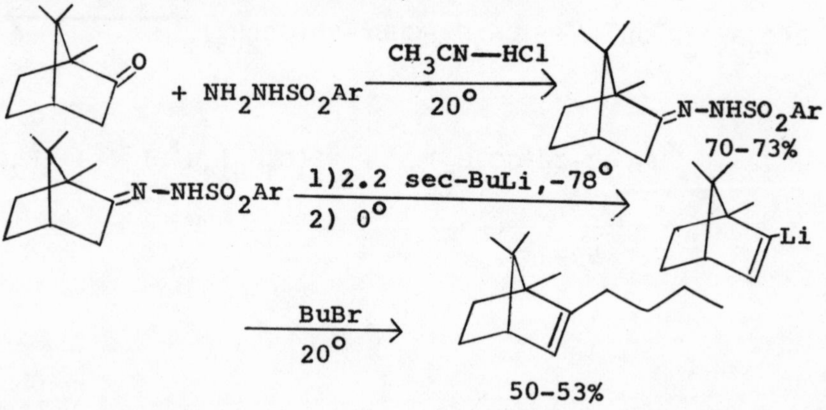

$$\text{camphor} + \text{NH}_2\text{NHSO}_2\text{Ar} \xrightarrow[20^\circ]{\text{CH}_3\text{CN——HCl}} \text{=N-NHSO}_2\text{Ar}$$

70-73%

$$\text{=N-NHSO}_2\text{Ar} \xrightarrow[2)~0^\circ]{1)2.2~\text{sec-BuLi},-78^\circ} \text{(vinyllithium)···Li}$$

$$\xrightarrow[20^\circ]{\text{BuBr}} \text{2-butylbornene}$$

50-53%

ε-Benzoylamino-α-chlorocaproic Acid

Y. Ogata, T. Sugimoto, and M. Inaishi, Department of
Applied Chemistry, Nagoya University, Furocho,
Chikusa-ku, Nagoya, Japan

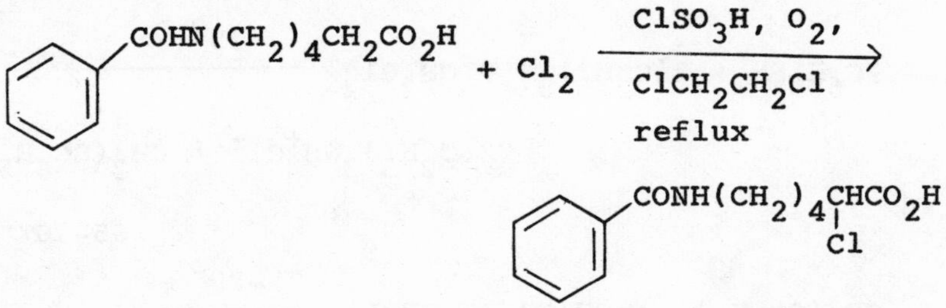

$$\text{COHN(CH}_2)_4\text{CH}_2\text{CO}_2\text{H} + \text{Cl}_2 \xrightarrow[\substack{\text{ClCH}_2\text{CH}_2\text{Cl} \\ \text{reflux}}]{\text{ClSO}_3\text{H, O}_2,} \text{CONH(CH}_2)_4\overset{|}{\underset{\text{Cl}}{\text{CH}}}\text{CO}_2\text{H}$$

80%

Glutaconaldehyde

J. Becher, Department of Chemistry, Odense University,
Campusvej 55, 5230 Odense M, DK-5000 Odense, Denmark

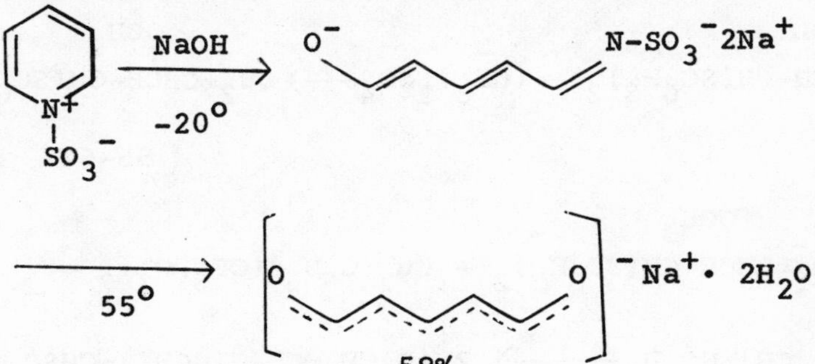

$$\xrightarrow[-20^\circ]{\text{NaOH}} \quad O^- \diagup\diagdown\diagup\diagdown\diagup\diagdown N\text{-}SO_3^-2Na^+$$

$$\xrightarrow{55^\circ} \quad \left[O \diagup\diagdown\diagup\diagdown\diagup\diagdown O \right]^- Na^+ \cdot\ 2H_2O$$

58%

2075 (*Z*)-1-Phenylthio-2-methoxy-1,3-butadiene

T. Cohen, R. J. Ruffner, D. W. Shull, E. R. Fogel, and J. R. Falck, Department of Chemistry, University of Pittsburg, Pittsburg, PA 15260

A. $$2C_6H_5SH + 2(C_2H_5)_3N + CH_2Cl_2 \xrightarrow{25^\circ} 2(C_2H_5)_3NH^+Cl^- + CH_2(SC_6H_5)_2$$

65–70%

B. $$CH_2(SC_6H_5)_2 + CH_3CH_2CH_2CH_2Li \xrightarrow{-20^\circ} LiCH(SC_6H_5)_2 + CH_3CH_2CH_2CH_3$$

$$LiCH(SC_6H_5)_2 + CH_2{=}CHCHO \longrightarrow CH_2{=}CH\overset{\displaystyle OLi}{\overset{|}{C}}H{-}CH(SC_6H_5)_2$$

$$CH_2{=}CH\overset{\displaystyle OLi}{\overset{|}{C}}H{-}CH(SC_6H_5)_2 + (CH_3)_2SO_4 \longrightarrow CH_2{=}CH\overset{\displaystyle OCH_3}{\overset{|}{C}}H{-}CH(SC_6H_5)_2$$

55–64%

C. $$2CH_2{=}CH\overset{\displaystyle OCH_3}{\overset{|}{C}}H{-}CH(SC_6H_5)_2 + Cu_2(C_6H_6)(CF_3SO_3)_2 +$$

$$2[(CH_3)_2CH]_2NC_2H_5 \longrightarrow 2\ (Z)CH_2{=}CH{-}C(OCH_3){=}CHSC_6H_5$$

70–85%

$$+\ 2CuSC_6H_5 + 2[(CH_3)_2CH]_2\overset{\displaystyle H}{\overset{|}{\underset{+}{N}}}C_2H_5\ \ 2CF_3SO_3^-$$

2077[a] **1-N-Acylamino-1,3-dienes: Benzyl *trans*-1,3-butadiene-1-carbamate**

P. J. Jessup, C. B. Petty, J. Roos, and L. E. Overman, Department of Chemistry, University of California, Irvine, Irvine, CA 92717

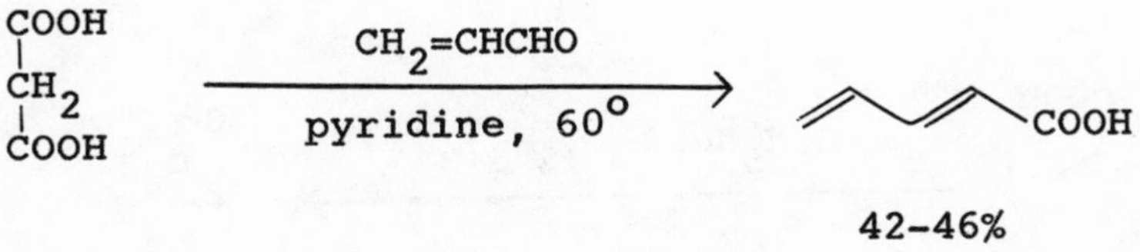
COOH
CH2
COOH
CH2=CHCHO
pyridine, 60°
COOH
42-46%

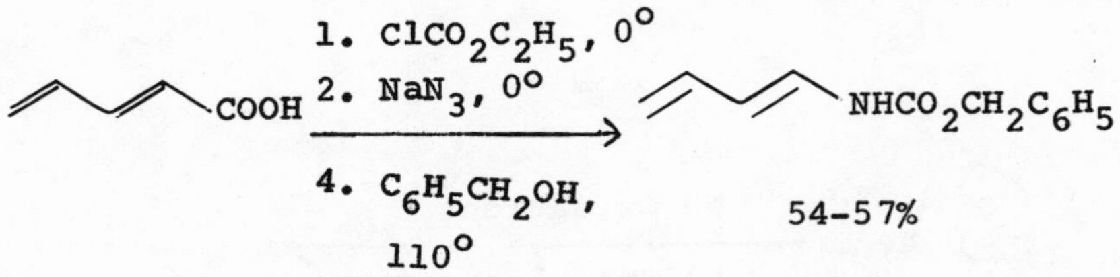
1. ClCO2C2H5, 0°
2. NaN3, 0°
4. C6H5CH2OH, 110°
COOH
NHCO2CH2C6H5
54-57%

2079[a] **4-(3-Pyridyl)-4-oxobutyronitrile**

H. Stetter, H. Kuhlmann, and G. Lorenz, Institut für Organische Chemie der Rheinisch-Westfälischen Technischen Hochschule, Aachen, Germany

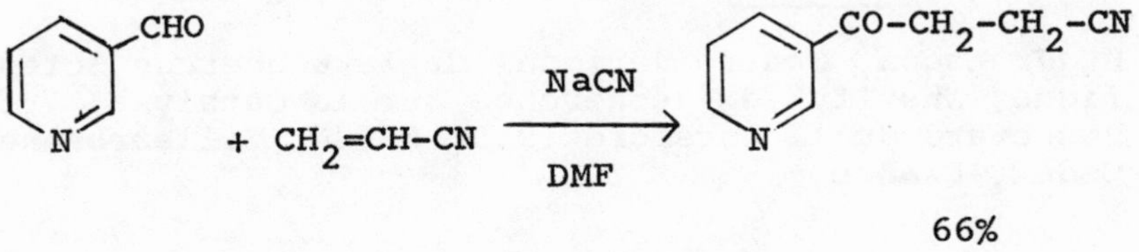
CHO
N
+ CH2=CH-CN
NaCN
DMF
CO-CH2-CH2-CN
N
66%

2081 **Phenylselenolactonization: Preparation of (1α,4α,5α)-4-(Phenylseleno)-6-oxabicyclo [3.2.2]nonan-7-one and 6-Oxabicyclo[3.2.2] non-3-en-7-one**

K. C. Nicolaou, Z. Lysenko, and S. Seitz, Department of Chemistry, University of Pennsylvania, Philadelphia PA 19104

A.

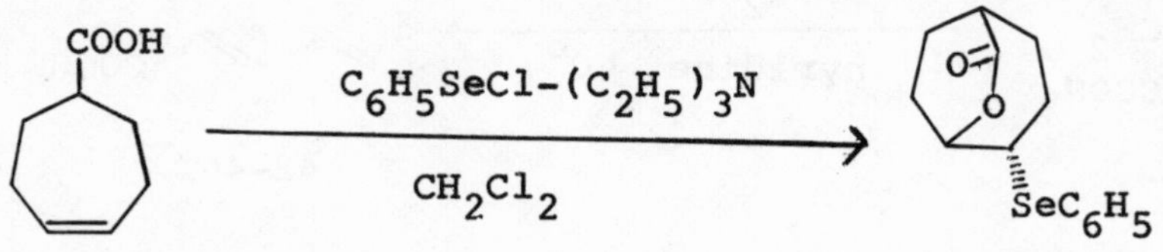

$$C_6H_5SeCl-(C_2H_5)_3N$$
$$CH_2Cl_2$$

99%

B.

a) Oxidation

b) <u>Syn</u>-elimination

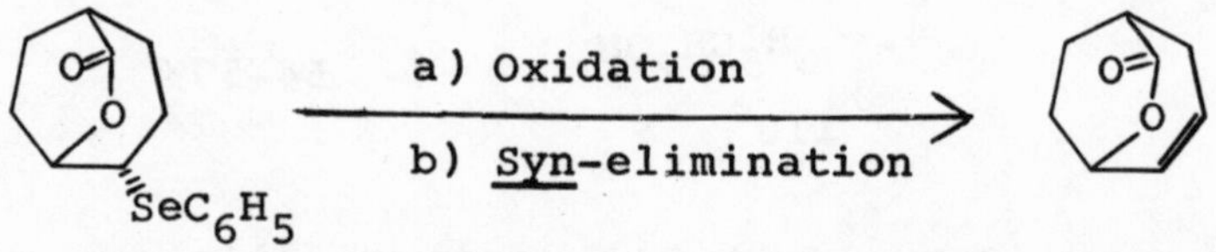

80%

2085 **The Diazo Transfer Reaction Under Phase Transfer Conditions: Di-<u>tert</u>-butyl Diazomalonate**

H. J. Ledon, Centre National de la Recherche Scientifique, Institut de Recherches sur la Catalyse, 79, Boulevard du 11 Novembre 1918, 69626, Villeurbanne Cedex, France

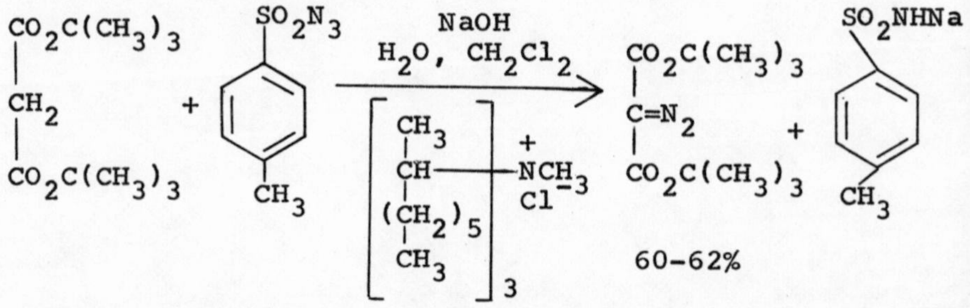

$CO_2C(CH_3)_3$ — CH_2 — $CO_2C(CH_3)_3$ + (tosyl azide, SO_2N_3, CH_3)

$$\xrightarrow[\substack{H_2O, CH_2Cl_2 \\ \left[\begin{array}{c} CH_3 \\ CH-NCH_3 \\ (CH_2)_5 \quad Cl^- \\ CH_3 \end{array}\right]_3}]{NaOH}$$

$CO_2C(CH_3)_3$ — $C=N_2$ — $CO_2C(CH_3)_3$ + (SO_2NHNa, CH_3)

60–62%

2087   Cis-Oxabicyclo[6.1.0]nonane

R. D. Bach and J. W. Knight, Department of Chemistry,
Wayne State University, Detroit, MI 48202

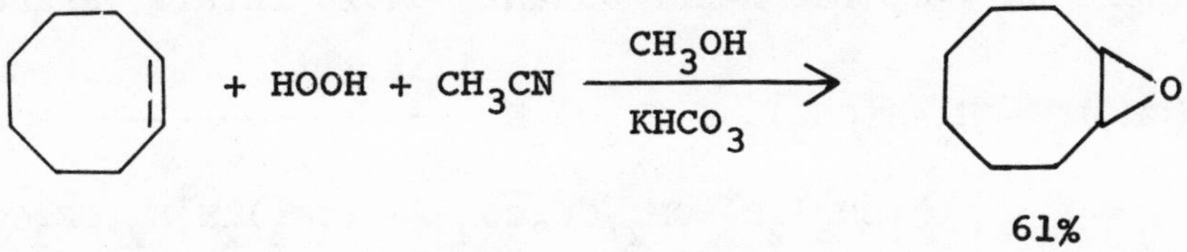

$$\text{(cyclooctene)} + HOOH + CH_3CN \xrightarrow[KHCO_3]{CH_3OH} \text{(cis-oxabicyclo[6.1.0]nonane)}$$

61%

2089   Reductive Deoxygenation via N,N,N',N'-Tetra-
methylphosphorodiamidate: 17β-Methoxy-5α-
2-androstene

R. E. Ireland, T. H. O'Neill, and G. L. Tolman, Division of Chemistry & Chemical Engineering, The Chemical Laboratories, California Institute of Technology,
Pasadena, CA 91125

$$POCl_3 + (CH_3)_2NH \xrightarrow{(C_2H_5)_2O} [(CH_3)_2N]_2POCl$$

74-84%

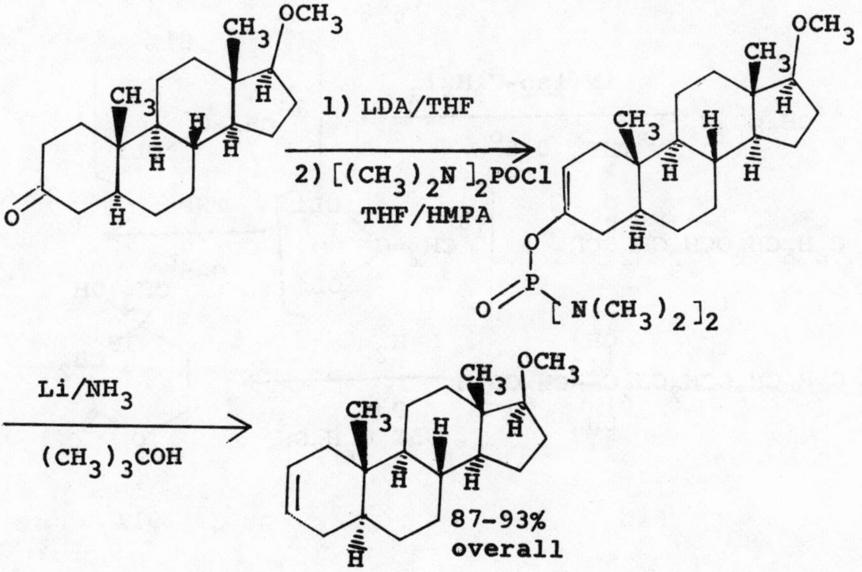

1) LDA/THF

2) [(CH$_3$)$_2$N]$_2$POCl
THF/HMPA

Li/NH$_3$

(CH$_3$)$_3$COH

87-93%
overall

1-Dimethylamino-4-methyl-3-pentanone

M. Gaudry, Y. Jasor, T. Bui Khac, and A. Marquet,
Centre National de la Recherche Scientifique, Centre
d'Etudes et de Recherches de Chimie Organique
Appliqué 2-8, rue Henri-Dunant, 94320 Thiais, France

$$(CH_3)_2N-CH_2-N(CH_3)_2 + CF_3CO_2H \longrightarrow$$

$$(CH_3)_2N^+=CH_2, CF_3CO_2^- + (CH_3)_2N^+H_2, CF_3CO_2^-$$

$$(CH_3)_2CH-CO-CH_3 + (CH_3)_2N^+=CH_2, CF_3CO_2^- \longrightarrow$$

$$(CH_3)_2CH-CO-CH_2-CH_2-N(CH_3)_2$$

53%

2094   Mevalonolactone-2-^{13}C

M. Tanabe and R. H. Peters, Bio-Organic Chemistry
Department, SRI International, Menlo Park, CA 94025

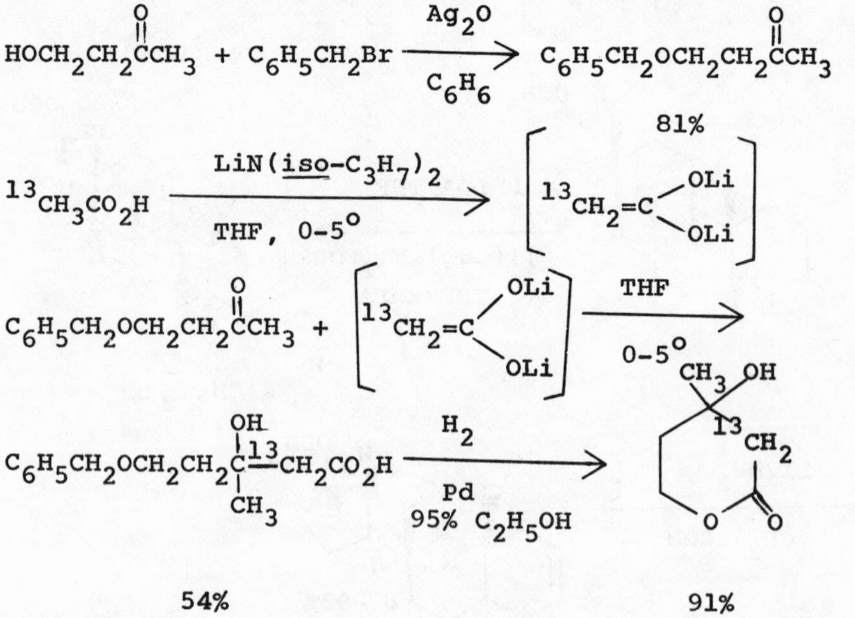

$$HOCH_2CH_2\overset{O}{\overset{\|}{C}}CH_3 + C_6H_5CH_2Br \xrightarrow[C_6H_6]{Ag_2O} C_6H_5CH_2OCH_2CH_2\overset{O}{\overset{\|}{C}}CH_3$$

81%

$$^{13}CH_3CO_2H \xrightarrow[THF, 0-5°]{LiN(\underline{iso}-C_3H_7)_2} \left[^{13}CH_2=C\begin{smallmatrix} OLi \\ \\ OLi \end{smallmatrix} \right]$$

$$C_6H_5CH_2OCH_2CH_2\overset{O}{\overset{\|}{C}}CH_3 + \left[^{13}CH_2=C\begin{smallmatrix} OLi \\ \\ OLi \end{smallmatrix} \right] \xrightarrow[0-5°]{THF}$$

$$C_6H_5CH_2OCH_2CH_2\underset{CH_3}{\overset{OH}{\underset{|}{\overset{|}{C}}}}{}^{13}CH_2CO_2H \xrightarrow[95\% \ C_2H_5OH]{H_2 \atop Pd}$$

54%

91%

Ethyl Thiazole-4-carboxylate

G. D. Hartman and L. M. Weinstock, Merck Sharp &
Dohme Research Laboratories, Division of Merck
and Company, Inc., West Point, PA 19486

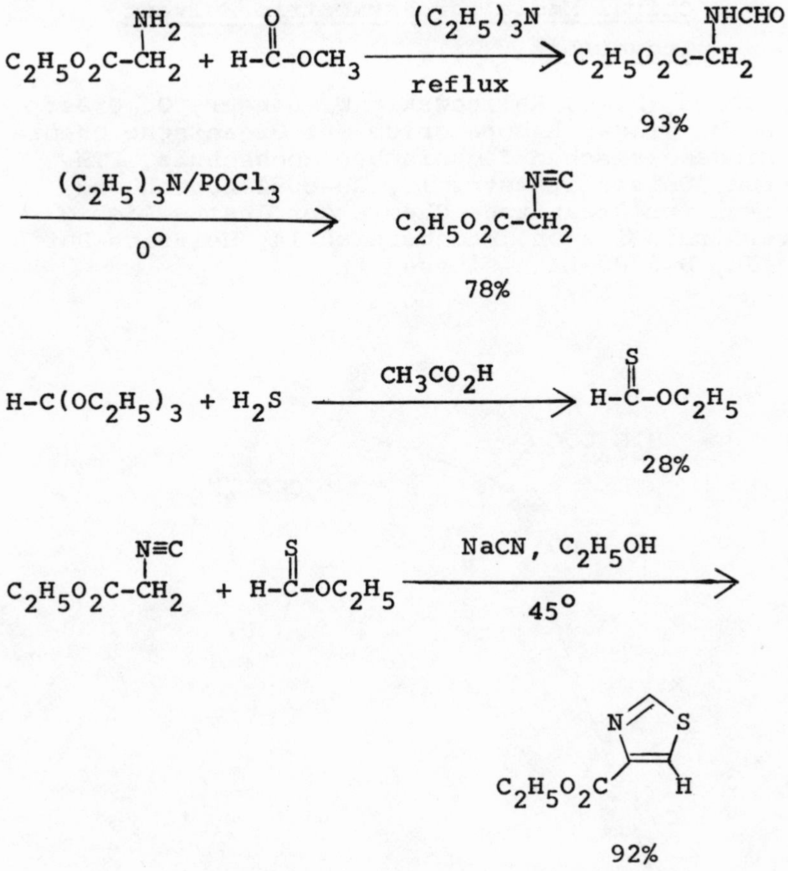

$$C_2H_5O_2C\text{-}\overset{\overset{NH_2}{|}}{C}H_2 \ + \ H\text{-}\overset{\overset{O}{\|}}{C}\text{-}OCH_3 \ \xrightarrow[\text{reflux}]{(C_2H_5)_3N} \ C_2H_5O_2C\text{-}\overset{\overset{NHCHO}{|}}{C}H_2$$

93%

$$\xrightarrow[0^\circ]{(C_2H_5)_3N/POCl_3} \ C_2H_5O_2C\text{-}\overset{\overset{N\equiv C}{|}}{C}H_2$$

78%

$$H\text{-}C(OC_2H_5)_3 \ + \ H_2S \ \xrightarrow{CH_3CO_2H} \ H\text{-}\overset{\overset{S}{\|}}{C}\text{-}OC_2H_5$$

28%

$$C_2H_5O_2C\text{-}\overset{\overset{N\equiv C}{|}}{C}H_2 \ + \ H\text{-}\overset{\overset{S}{\|}}{C}\text{-}OC_2H_5 \ \xrightarrow[45^\circ]{NaCN, \ C_2H_5OH}$$

$$C_2H_5O_2C\text{-}\underset{H}{\overset{N}{\diagdown}}\text{thiazole}$$

92%

<u>(S.S)-(+)-1,4-Bis(dimethylamino)-2,3-dimethoxybutane (DDB) and (S.S)-(-)-1,2,3,4-Tetramethoxybutane (TMB) from R.R-(+)-Tartaric Acid Diethyl Ester:</u> <u>Chiral Media for Asymmetric Solvent Inductions</u>

D. Seebach, H.-O. Kalinowski, W. Langer, G. Crass, and E.-M. Wilka, Laboratorium für Organische Chemie der Eidgenössischen Technischen Hochschule, ETH-Zentrum, Universitätstr. 16, CH-8092-Zurich and Institut für Organische Chemie der Justus Liebig-Universität Giessen, Fachbereich 14, Heinrich-Buff-Ring 58, D-6300-Lahn-Giessen

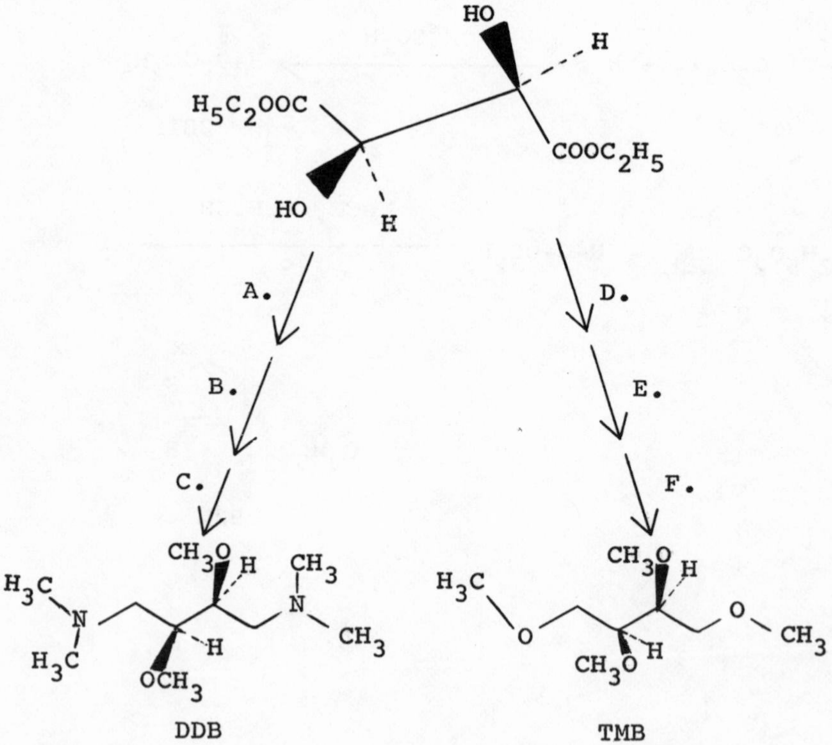

DDB

TMB

A. $H_5C_2OOC-\overset{\displaystyle OH}{\underset{\displaystyle OH}{CH}}-CH-COOC_2H_5$ $\xrightarrow{\text{HN(CH}_3)_2}$

$(H_3C)_2N-CO-\overset{\displaystyle OH}{\underset{\displaystyle OH}{CH}}-CH-CO-N(CH_3)_2$

93-95%

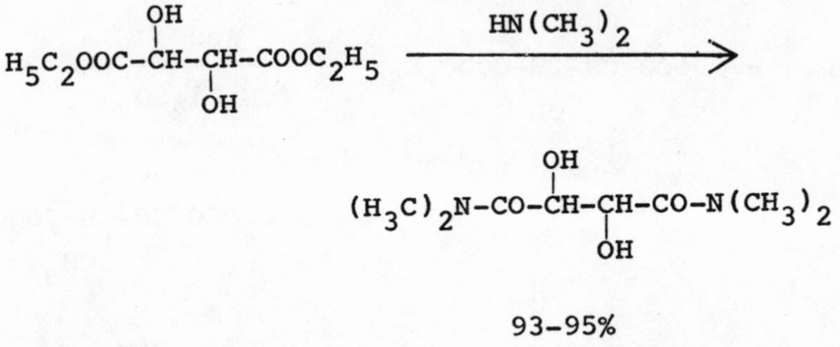

B. $(H_3C)_2N-CO-\overset{\displaystyle OH}{\underset{\displaystyle OH}{CH}}-CH-CO-N(CH_3)_2$ $\xrightarrow[\substack{NaOH \\ TEBA}]{(CH_3O)_2SO_2}$

$(H_3C)_2N-CO-\overset{\displaystyle OCH_3}{\underset{\displaystyle OCH_3}{CH}}-CH-CO-N(CH_3)_2$

95%

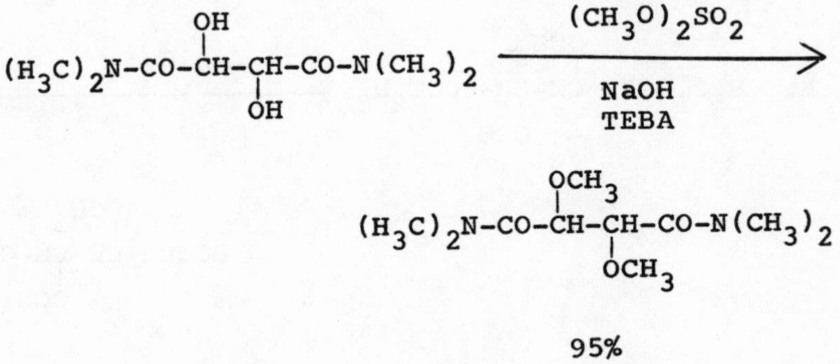

C. $(H_3C)_2N-CO-\overset{\displaystyle OCH_3}{\underset{\displaystyle OCH_3}{CH}}-CH-CO-N(CH_3)_2$ $\xrightarrow{\text{LiAlH}_4}$

$(H_3C)_2N-CH_2-\overset{\displaystyle OCH_3}{\underset{\displaystyle OCH_3}{CH}}-CH-CH_2-N(CH_3)_2$

DDB

88%

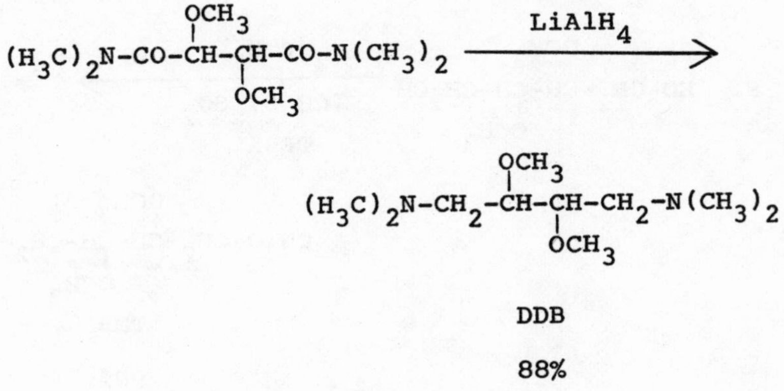

D. $H_5C_2OCO-\overset{\overset{\displaystyle OH}{|}}{CH}-\underset{\underset{\displaystyle OH}{|}}{CH}-COOC_2H_5$ $\xrightarrow[\text{(CH}_3\text{O)}_2\text{SO}_2]{\text{NaH}}$

$H_5C_2OCO-\overset{\overset{\displaystyle OCH_3}{|}}{CH}-\underset{\underset{\displaystyle OCH_3}{|}}{CH}-COOC_2H_5$

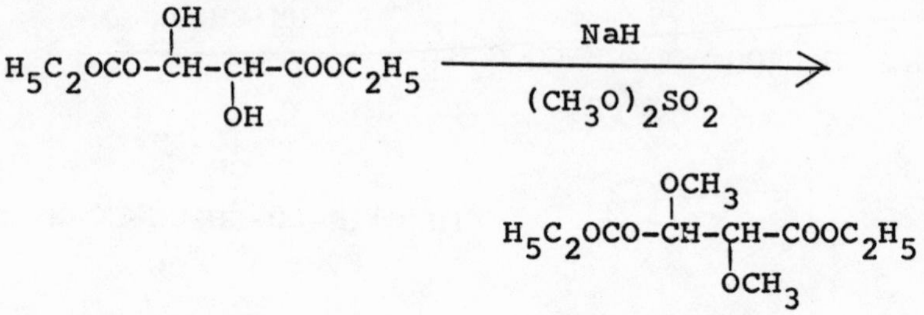

91%

E. $H_5C_2O-CO-\overset{\overset{\displaystyle OCH_3}{|}}{CH}-\underset{\underset{\displaystyle OCH_3}{|}}{CH}-COOC_2H_5$ $\xrightarrow{\text{LiAlH}_4}$

$HOCH_2-\overset{\overset{\displaystyle OCH_3}{|}}{CH}-\underset{\underset{\displaystyle OCH_3}{|}}{CH}-CH_2OH$

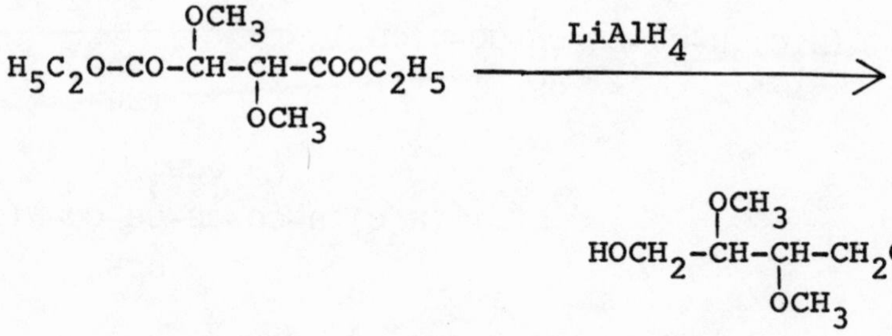

71-75%

F. $HO-CH_2-\overset{\overset{\displaystyle OCH_3}{|}}{CH}-\underset{\underset{\displaystyle OCH_3}{|}}{CH}-CH_2OH$ $\xrightarrow[\substack{\text{(CH}_3\text{O)}_2\text{SO}_2 \\ \text{TEBA}}]{\text{NaOH}}$

$CH_3O-CH_2-\overset{\overset{\displaystyle OCH_3}{|}}{CH}-\underset{\underset{\displaystyle OCH_3}{|}}{CH}-CH_2-OCH_3$

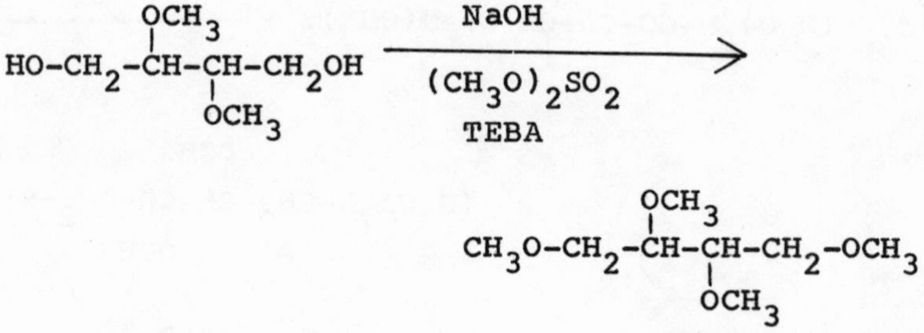

TMB

90%

5,5-Diethoxy-1,3-cyclopentadiene and Its Dimer, 5,5,10,10-Tetraethoxytricyclo [5.2.1.0^{2,6}]deca-3,8-diene

J. C. Barborak, Department of Chemistry, University of North Carolina at Greensboro, Greensboro, NC 27412

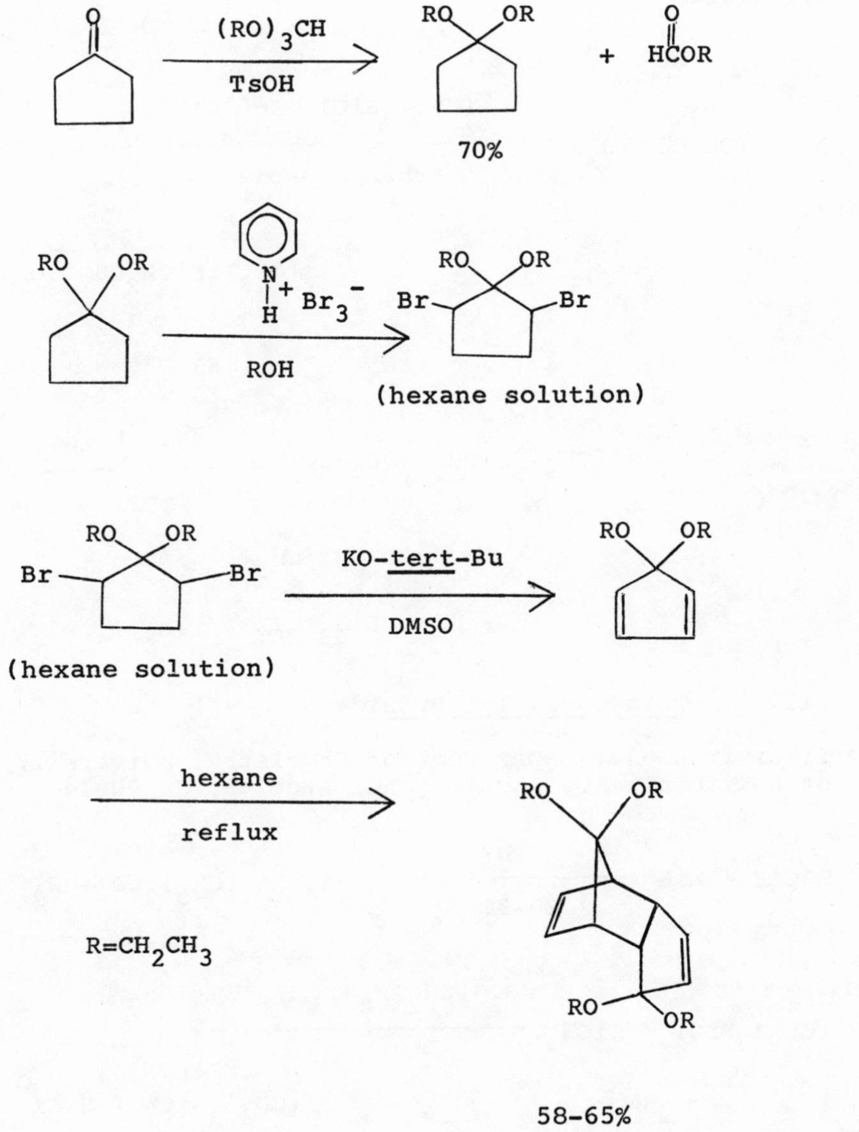

$$R = CH_2CH_3$$

Silylation of Ketones with Ethyl Trimethyl-
          silylacetate: (Z)-3-Trimethylsiloxy-2-
          pentene

I. Kuwajima, E. Nakamura, and K. Hashimoto, Department
of Chemistry, Tokyo Institute of Technology, Tokyo
152, Japan

A. $BrCH_2COOC_2H_5$ $\xrightarrow[\text{ether/benzene}]{(CH_3)_3SiCl,\ Zn(Cu)}$

$(CH_3)_3SiCH_2COOC_2H_5$

63-74%

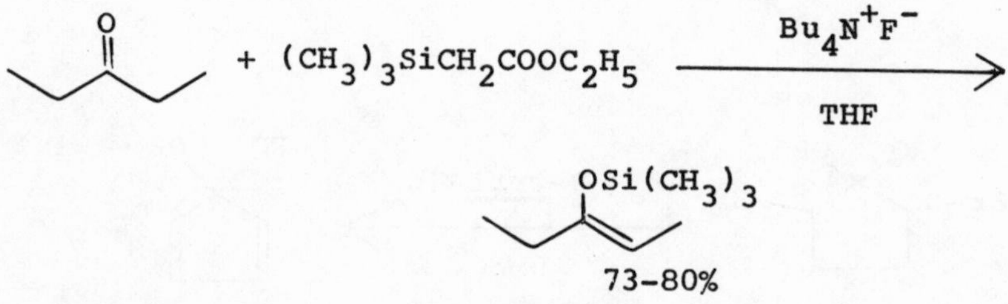

+ $(CH_3)_3SiCH_2COOC_2H_5$ $\xrightarrow[\text{THF}]{Bu_4N^+F^-}$

$\overset{OSi(CH_3)_3}{}$

73-80%

2110   Trimethylsilyl Cyanide

T. Livinghouse, Department of Chemistry, University
of California Los Angeles, Los Angeles, CA 90024

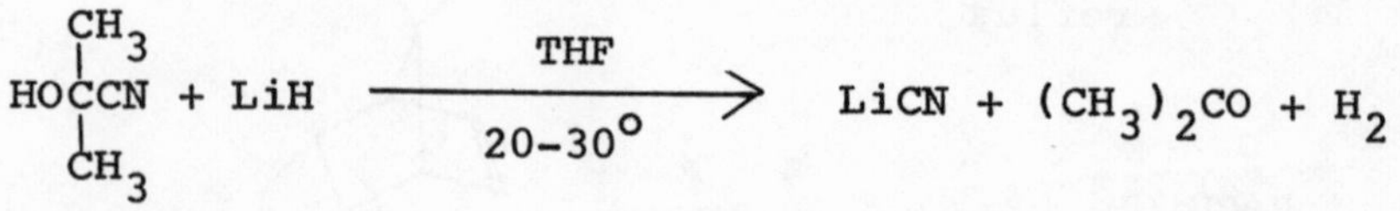

$\overset{CH_3}{\underset{CH_3}{HOCCN}}$ + LiH $\xrightarrow[20-30^\circ]{THF}$ LiCN + $(CH_3)_2CO$ + H_2

$(CH_3)_3SiCl$ + LiCN $\xrightarrow[25^\circ]{CH_3O(CH_2CH_2O)_4CH_3}$

$(CH_3)_3SiCN$ + LiCl

59-82%

2111 Osmium-Catalyzed Vicinal Oxyamination of Olefins by Chloramine-T: *Cis*-2-(*p*-Toluenesulfonamido)cyclohexanol and 2-Phenyl-1-(*p*-Toluenesulfonamido)propan-2-ol

E. Herranz and K. B. Sharpless, Department of Chemistry, Stanford University, Stanford, CA 94305

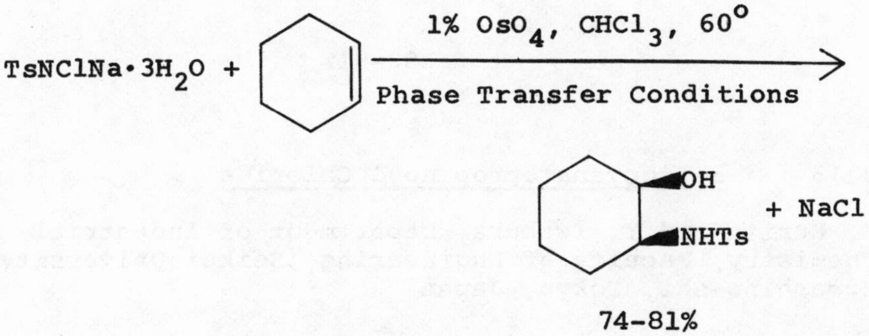
TsNClNa•3H2O +
1% OsO4, CHCl3, 60°
Phase Transfer Conditions
OH
NHTs
+ NaCl
74-81%

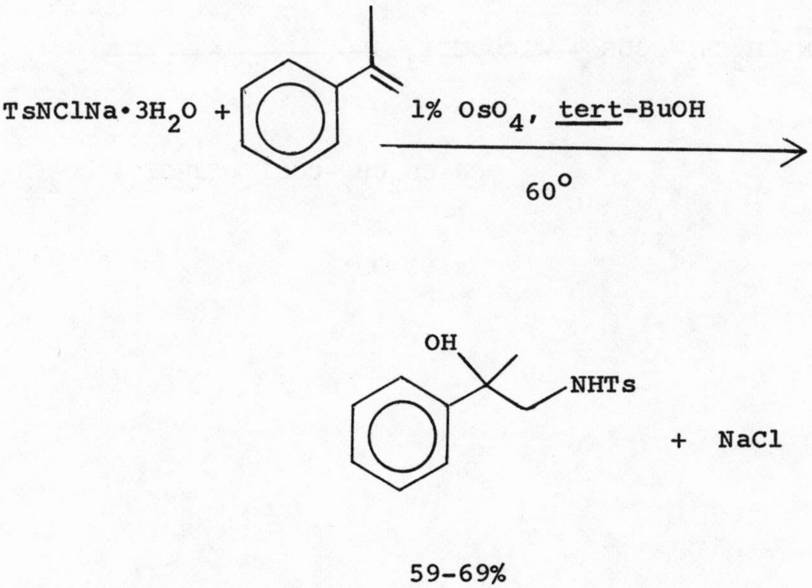
TsNClNa•3H2O +
1% OsO4, tert-BuOH
60°
OH
NHTs
+ NaCl
59-69%

2112   <u>Trichloromethyl Chloroformate</u>

(Diphosgene)

K. Kurita and Y. Iwakura, Department of Industrial
Chemistry, Faculty of Engineering, Seikei University,
Musashino-shi, Tokyo, Japan

$$ClCOOCH_3 + 3Cl \longrightarrow ClCOOCCl_3 + 3HCl$$

82-91%

2113   <u>3-Isocyanatopropanoyl Chloride</u>

K. Kurita and Y. Iwakura, Department of Industrial
Chemistry, Faculty of Engineering, Seikei University,
Musashino-shi, Tokyo, Japan

$$H_2N-CH_2CH_2-COOH + ClCOOCCl_3 \longrightarrow$$

$$OCN-CH_2CH_2-COCl + 3HCl + CO_2$$

93-97%